# *Ladybirds*
## *of Surrey*

# *Ladybirds*

## *of Surrey*

ROGER D. HAWKINS

SURREY WILDLIFE TRUST

Cover illustration
Orange Ladybird *(Halyzia 16-guttata)* by David Element

ISBN 0 9526065 5 0

British Library Cataloguing-in-Publication Data.
A catalogue record for this book is available
from the British Library.



First published 2000
by Surrey Wildlife Trust
School Lane, Pirbright, Woking, Surrey GU24 0JN.

Printed and bound in Great Britain by
Biddles Ltd, Guildford and King's Lynn

# FOREWORD

A well-known advertising slogan runs: 'Everyone loves a ladybird'. In many parts of the world, ladybirds are seen as foretellers of love. These connections between ladybirds and love come to mind because *Ladybirds of Surrey* is indeed a labour of love. It is the fruit of 20 years of painstaking recording and observation undertaken by Roger Hawkins and his dedicated group.

*Ladybirds of Surrey* is so much more than a set of distribution maps showing which ladybirds may be found where and on what in Surrey. In fact, the rest of the book is as pertinent to a ladybird observer in John o' Groats or Land's End as it is to a naturalist in Epsom.

It provides an excellent introduction to identification of these brightly coloured beetles, with high quality photographs of both larvae and adult ladybirds covering all the common forms of the more variable species. The identity of most ladybirds found should thus be discernible by simply referring to the photographs.

The whole book is packed with useful details, written in a very readable style devoid of jargon and scientific gobbledegook. We read fascinating facts about ladybirds in general, as well as in-depth accounts of the distinguishing features, distribution, host plants and habits of each species. These accounts are drawn from the detailed diary notes of the recorders and provide an intimate insight into the lives of ladybirds that has not been achieved in any of the more general literature on the subject.

The untold hours of labour that are devoted to guides of this type are a particular feature of the British. This has given us the richest tradition of wildlife recording on Earth. It has also perhaps contributed to the view that others have of some of us as being a little eccentric. If so, long may that view continue. It is only through the dedicated work, given freely by people such as those who have contributed to this book, that our knowledge and understanding of the fauna and flora of the British Isles maintains its depth and diversity.

I am thus honoured to write this foreword and have the chance to publicly thank Roger Hawkins for the superb publication that he and his team have produced.

MICHAEL MAJERUS

# CONTENTS

# PREFACE

Although the main purpose of this book is to report on a 20-year-long survey into the distribution and habits of ladybirds in the county of Surrey, the reader who knows little about these insects is encouraged to become more expert by the inclusion of considerably more detailed text on identification than in earlier books in this series.

In Britain there are just over 40 different species of ladybird, the name we give to beetles of the family Coccinellidae. About 24 of the larger, more typical species are brightly coloured and have a pattern of spots. They all have English names and most can be named on sight, or in the most difficult cases by checking a few simple features with a pocket lens. The basic colour may be red, orange, yellow, black or creamy-white, while the spots themselves may be black, red or white. The number of spots varies from one to 24, with one or more species to almost every number, although not all of them occur in Britain.

All these larger species can be named accurately using this book. For many of them, a search through the colour plates will be sufficient, for they have a unique colour pattern and number of spots. A few species are variable, some extremely so, and the more important varieties are also illustrated. The text about each species lists the additional features to be checked in order to confirm the identification.

The family includes about 18 smaller ladybirds whose identification is more difficult and beyond the scope of this book, although it can often be guessed from the size, appearance and habitat of the insect. These species are known only by their scientific names, are small or very small and are not always clearly recognisable as ladybirds. We cover their distribution and habits as fully as possible, but not their identification and only a few of them are illustrated.

Ladybird larvae are little-known but are just as varied and almost as colourful as the adults. Their bodies are covered with projections which vary from low bumps with short hairs to tall bristly spines. All the larvae can be identified through their particular arrangement of hairs and spines, but many of them also have a unique colour pattern. Twenty different species are illustrated in this book and about 15 of them can be named merely through examining the pattern of spots on the body, as explained in the text for each species.

The separate accounts in this book of the different species of ladybirds are largely based on observations made in Surrey during the survey, based either on my own ladybird diary or on similar documents kept by other recorders. For 20 years I have written down a note about almost every ladybird seen, recording its exact location, what it was doing and how many there were together. About six other people have also supplied a field note with every record, while the remainder have restricted such observations to the rarer species and anything out of the ordinary.

One result of this systematic approach is that far more information has been collected about the common species than about the rare ones, so the amount written on each species is in proportion to its abundance, unlike those books on birds that dismiss the wren in a few lines while devoting page after page to the Dartford warbler. One effect of detailed general observation is that several questions are raised for every one that is answered. I would be gratified indeed if any idea coming out of this survey is selected as a suitable subject for future research.

My intention at the start of the survey was to investigate the life-history and biology of each species as well as its distribution, and in particular to work out the nature of its food. With the possible exception of the Pine Ladybird on broad-leaved trees, this ambitious plan has not been fully realised, for three reasons. Firstly, the advent of the Cambridge Ladybird Survey led to much of this ground being covered on a national basis. Secondly, much of the information was already available; in particular, the valuable but little-known paper by Mills (1981) lists the prey items of many species of ladybird in the wild in Britain, both as adults and larvae. Thirdly, my ability to name aphids in the field is still negligible. This is not as difficult as it first appears, since many aphids are specialised in their choice of host-plant, but some critical examination of specimens is necessary, using microscopical techniques. A useful starting-point is *Aphid Predators* by Rotheray (1989).

An explanation is needed for the Molesey phenomenon. Three out of the four ladybirds found new to Surrey during this survey occurred together in and around a single garden in West Molesey in 1997. These were the large plant-feeding Bryony Ladybird found for the first time in Britain, the small brown *Rhyzobius chrysomeloides* found only for the second time, and the tiny four-spotted *Nephus quadrimaculatus*, long considered a national rarity. The habitat was nothing special – an ordinary suburban garden with no particular features or reason why these three rare ladybirds should be living there together. It seems rather that attention was attracted to the area by the presence of the Bryony Ladybird, which must have had its initial British colony in Molesey or somewhere nearby, and this led to the two smaller species also being found there, although they were probably already widespread in the county.

# ABOUT THE SURVEY

Surveys of insect groups in Surrey owe their origin to the botanical surveys of the 1960's in which the detailed distribution of wild plants was shown as dot maps on a 2-kilometre grid. In our case the inspiration was *Flora of Surrey* by J.E. Lousley (1976). Arriving in Surrey as a young amateur botanist, I could find nowhere where the recorders had not been and every species map showed where the plants really occurred. David Baldock helped with this botanical survey and later made pioneering maps of insects which contributed to *Dragonflies of Surrey* and, after almost 30 years recording, culminated in *Grasshoppers and Crickets of Surrey*. The mapping of ladybirds followed on from these and so is a direct descendant of the botanical work. Recording began tentatively in 1980 and as a full-scale project from 1984 onwards.

Most recording of insects, whether done on a casual basis or organised with some definite objective in mind, takes place on the best sites or represents a search for the rarest species. The recording of ladybirds in Surrey is no exception and our favoured sites include the heathy commons of the north and west of the county, some fine remaining stretches of chalk downland, and Bookham and Ashtead Commons with their long history of recording and management. Many of our recorders have made lists of ladybirds for such places, but distribution maps created on this basis show only a thin scatter of records within the grid of over 500 tetrads (2 km by 2 km squares) and the maps for common species look no different from the maps for rare ones.

My own part in this survey has been to record those areas of indifferent habitat, the suburban and agricultural areas that occupy most of the county. This has been approached in a systematic way by visiting each tetrad in turn and aiming to record a minimum number of species, initially seven in the built-up areas and ten in the countryside. The map overleaf shows to what extent this has been achieved. All corners of the county, whether easily accessible or remote, have been reached by using our excellent network of public transport and then on foot, so the accounts and maps record largely the distribution of ladybirds as found along public footpaths. In Surrey these are almost always open, well-used and in good condition, and both county and local councils are to be congratulated on this. The only serious path-blocker is the Department of Transport, which appears to treat both the safety and legal rights of pedestrians with an attitude bordering on contempt by upgrading roads such as the A3 to carry traffic at a density and speed of motorway standards, even though the road is crossed by public rights-of-way on the level.

Our survey area includes those parts of former Surrey now included in Greater

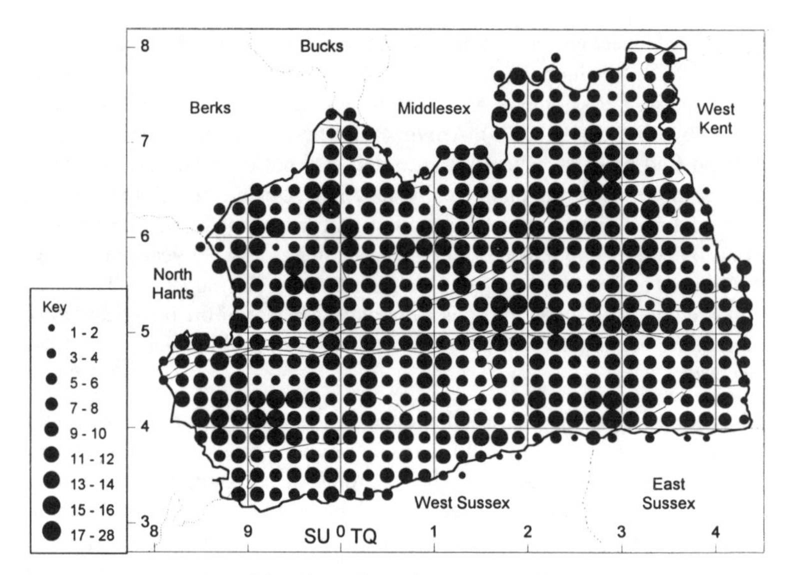


***Number of species per tetrad***

London and extends right up to the River Thames in central London. The built-up areas of inner London have been visited just as regularly as the countryside. Contrary to the impression sometimes created by stories in newspapers, I found south London to be a civilised place inhabited by happy, friendly people. The boroughs of Lambeth and Southwark in particular deserve our thanks for their many fine parks and commons which often have small areas set aside for wildlife. The London Wildlife Trust has been a major influence in the conservation of such areas which have provided significant numbers of records for our distribution maps.

The densely-populated London region lies in the north-eastern part of our survey area, while the south-western part is comparatively rural and sparsely-populated. The ideas for recording insects in Surrey and producing this series of books had their origin in discussions at the entomology section of the Croydon Natural History and Scientific Society, so our original team of recorders were all based in east Surrey within easy reach of Croydon. Even after further recruitment, at least 80% of our active recorders live in the east of the county. To counteract this bias towards the east, very many journeys to west Surrey have been made by train and car, but some vestige of bias towards east Surrey remains in the maps for a few species.

One sets out on a project such as this with great enthusiasm, for the detailed mapping of any group of creatures for the first time is a voyage into the unknown, and the first few dots on maps for the more familiar species arouse great interest. As the years go by and the distribution maps fill up, a sense of disillusionment begins to set in, for most species appear to be found everywhere. It also becomes clear that surveying even a small group over a limited area takes up a substantial slice of a life-time when done single-handedly or by only a few people. Having persevered to the point where the distribution maps are sufficiently complete to allow publication, a look back at the positive side of detailed distributional mapping can be made concerning ladybirds in Surrey.

The abundance or rarity of each species can be seen at a glance from its distribution map. For instance, species of the same habitat (such as pine trees or grassland) can be ranked, with those that are widespread and abundant separated from those of more limited distribution. Our maps are now complete enough to begin to show that most difficult of distributions to portray, the species that occurs almost everywhere. There have been some other unexpected developments. Two European species of ladybird have arrived in Britain and apparently established themselves in Surrey, and are increasing (one of these is small and insignificant but the other is the large and distinctive Bryony Ladybird). Two national rarities, both small species, have been found for the first time in Surrey and one of them has spread over about half the county while the other remains confined to one site.

It is more surprising still that two species appear to have adapted to a new kind of food in recent years. The 24-spot Ladybird was recorded chiefly from red campion in the 1960's but now feeds principally on the false oat-grass, a more abundant plant, which must have allowed it to increase its numbers greatly. The additional food of the Pine Ladybird is clearly novel, since its new prey is an introduced species of scale-insect that has been known in the country for less than 40 years, but even so both prey and predator have built up a huge population in a new habitat for the ladybird. Another new discovery, that the larvae of the mildew-feeding Orange Ladybird are most common under the leaves of sycamore, was made by ladybird workers from outside the county although some of the first observations of this were made in Surrey.

This detailed local investigation of ladybirds in Surrey overlapped with the large-scale, national and highly successful Cambridge Ladybird Survey which ran from 1984 to 1994. This originated in the Department of Genetics of Cambridge University with investigations into the meaning of pattern variation in ladybirds. Work on the 2-spot Ladybird was extended to all species and each was reared and studied in the laboratory. This led to involving and encouraging large numbers of interested amateurs and enthusiastic children throughout the country to investigate the behaviour and national distribution of ladybirds. The result has been a succession of scientific papers and two fine books from the pens of Dr Michael Majerus and his students and colleagues.

Although records from Surrey have of course been included in the national distribution maps of the Cambridge Ladybird Survey, and some field work for the Cambridge research has taken place in our county, it has not been possible, even with the greatest goodwill on both sides, to extract the Surrey records from the enormous mass of correspondence held at Cambridge. However, we turn this deficiency into a virtue. On any matter concerning the food, abundance or life-history of any of the ladybird species, this book offers a completely independent second opinion which will in most cases support the conclusions derived from the Cambridge Survey.

Both the Cambridge Survey and this local one may have been prompted in part by the interest created in the population explosion of 7-spot Ladybirds that occurred during the summer drought of 1976. For the past 20 years, my detailed notes on ladybirds have been kept in the hope of providing an explanation as to how these occasional plagues of ladybirds are caused. At times it seemed that the 7-spot was quite abundant (and a population explosion was confidently predicted) but no massive increase in numbers has occurred. Any person now starting a similar ladybird diary for the 21st Century might find that the diary is recording the origins of another great ladybird year and explaining the factors that cause it.

Another conclusion from the survey is that the timing of the ladybird life-cycle varies from year to year and also differs between species. Even the number of generations may be different from one year to the next. Each species is different in its reaction to varying weather patterns, new opportunities for feeding and changes to its habitat. Some species are currently doing well and their numbers are increasing, while the population of others is decreasing or perhaps only fluctuating and temporarily at a low level. The detailed study of ladybirds in other parts of Britain would undoubtedly show some differences in the habits of the species from those described here, due to local variations in habitat or climate. But even in Surrey, a new ladybird diary kept for the next 15 to 20 years might well produce significantly different conclusions from this one, and probably in some unexpected way, such as the arrival of yet another new species or the movement of a native species to a new habitat or a new food.

# SURREY – THE SURVEY AREA

***Surrey in relation to southern England***

The Surrey of this survey is the traditional county rather than the modern one. It is the Surrey of the County Cricket Club and of the Boat Race, in which the north bank of the Thames is forever Middlesex and the south bank is forever Surrey. This traditional county is used for biological recording so that comparisons can be made between past, present and future surveys. It dates from the proposal by H.C. Watson in 1852 to divide Britain into units of approximately equal size for the purpose of botanical recording. He chose the county as his basic unit, but divided large counties into smaller units which he named vice-counties. Surrey was about the right size so became a vice-county without alteration, but larger counties such as Sussex, Kent and Hampshire were each divided into two vice-counties, while Yorkshire was split into five parts and small counties like Rutland were amalgamated with their neighbours. Modern counties have been set up to have roughly equal human populations, which is of little relevance to biological recording.

The Surrey of our survey, known as Watsonian vice-county 17, includes the following districts that are no longer part of the modern county of Surrey: the London boroughs of Wandsworth, Lambeth and Southwark that were transferred to the London County Council in 1889; the London boroughs of Richmond, Kingston, Merton, Sutton and Croydon that formed part of Greater London in 1965; Gatwick Airport, transferred to West Sussex in 1974.

The only significant part of modern Surrey outside our survey area is the borough of Spelthorne, including Staines, Ashford and Sunbury, which is north of the Thames and was transferred from Middlesex in 1965. The parish of Dockenfield to the south of Farnham is also part of modern Surrey although in the vice-county of North Hampshire.

In the north-east the boundary of the vice-county follows approximately that separating the London boroughs of Croydon and Southwark from Bromley and Lewisham, but also includes the former urban district of Penge and the grounds of the Crystal Palace that are part of modern Bromley. The vice-county boundary differs slightly from the modern boundary between the boroughs of Southwark and Lewisham and can only be followed by studying old maps.

The vice-county system is explained by Dandy (1969) who included maps at a scale of 1:625,000 showing the boundaries for the whole of Great Britain.

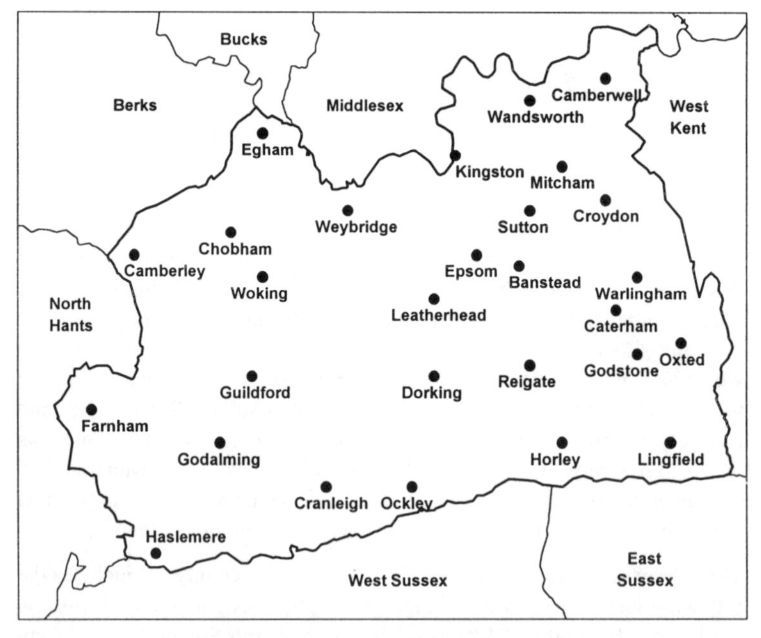


*Surrey in relation to bordering vice-counties*

# GEOLOGY, CLIMATE AND DISTRIBUTION

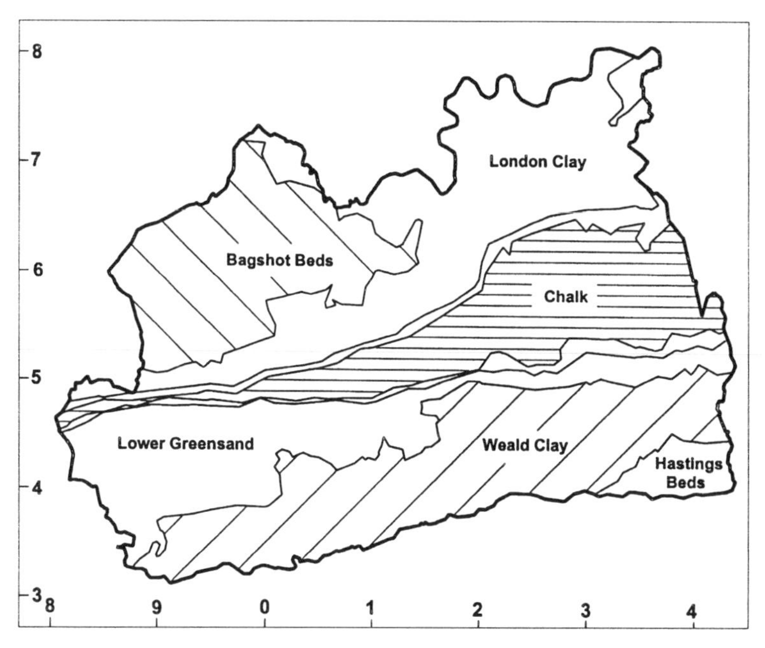


*Solid geology of Surrey*

As every cyclist knows, south-east England is crossed from east to west by parallel ranges of low hills which turn the simple matter of a bicycle ride from London to Brighton into a marathon race over hurdles. The principal ranges in Surrey are the Chalk, forming the North Downs with its steep south-facing escarpment, and the Lower Greensand. The Chalk is broader in the east where it is broken up by dry valleys and often overlain by clay, but the Greensand broadens out in the west and forms the extensive heathy commons around Thursley and Hindhead. There are further areas of sand in the Bagshot Beds in the north-west of the county and, to a very limited extent, in the Hastings Beds in the extreme south-east. The broad outline of the geology is shown on the map above and repeated on the distribution maps for each individual species.

The underlying geology affects the drainage and the nature of the soil, and therefore also the distribution of plants and the insects that depend on them. It is not so relevant for many of the more common ladybirds which are highly adaptable and little affected by geology, since most are predatory and can feed on different kinds of aphids or scale-insects. Many species are ubiquitous, while others are common in restricted habitats that occur widely throughout the county.

The Chalk is a dominant feature in Surrey and has many associated plants and insects. Three small, rare species of ladybirds (*Scymnus femoralis*, *Scymnus schmidti*, *Platynaspis luteorubra*) appear to be found predominantly on chalky soils but there are few records and the first two of these also occur in dry sandy areas.

The sandy soil of the Greensand and Bagshot Beds encourages the formation of heathland dominated by the common heather, which has both Heather and Hieroglyphic Ladybirds closely associated with it. Pine trees are abundant on these soils, both self-sown and in plantations, and these support almost a quarter of our ladybirds (Striped, Eyed, Cream-streaked, Pine, 18-spot, Larch and 10-spot Ladybirds, *Scymnus suturalis*, *Scymnus nigrinus*), many of which are found only on pine. Some of them are largely restricted to the major expanses of heath and pine in north and west Surrey, while others occur on planted pines throughout the county. The Scarce 7-spot Ladybird, always living near nests of the wood ant *Formica rufa*, is also found almost exclusively on sandy soils.

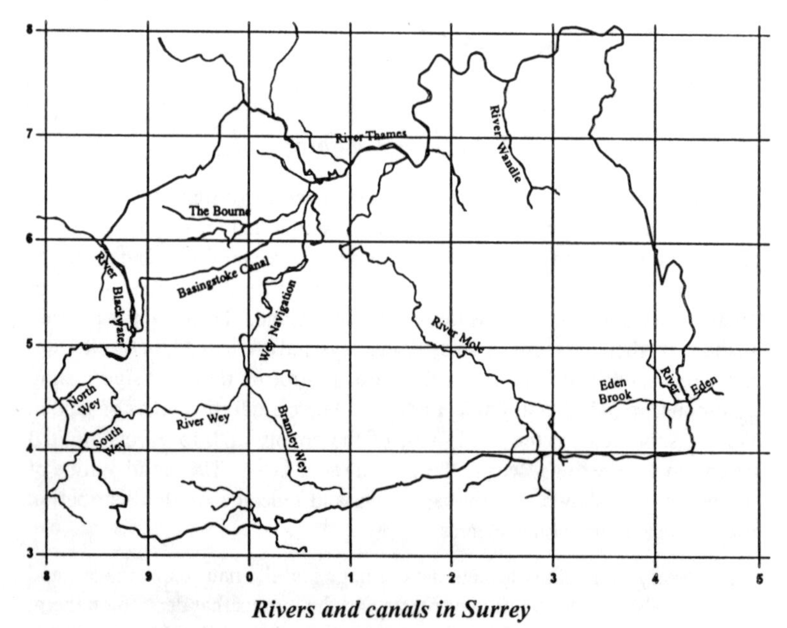


***Rivers and canals in Surrey***

The two main rivers of Surrey, the Wey and the Mole, flow approximately from south to north across the ranges of hills, having cut gaps in them over the ages. The county's wetlands, whether ponds, lakes or marshes, lie predominantly along these river valleys or on the low-lying flat land of the

Weald and London Clays. Three species of ladybirds are found in wet places or by the water's edge. Two of these are common (Water Ladybird, *Coccidula rufa*) while one is rare (*Coccidula scutellata*).

In common with other inland areas of south-east England, Surrey has a more continental climate than most of the British Isles, with warmer summers and colder, drier winters. Only one of the British ladybirds, the 5-spot, has a north-western type of distribution and is absent from Surrey. Many others are more abundant towards the south-east and we have all of them. The two new species recorded in the last five years indicate that we are also favourably placed to receive species from the Continent which are increasing their range towards the north.

Differences in climate exist even within the county, for the annual rainfall increases by almost half going south-west across Surrey from London towards the hilly area around Haslemere. This difference is accentuated by the warming effect of the built-up area of London, which may help to explain why our only permanent colonies of the Adonis Ladybird seem to be in the London area. The distribution of the 2-spot Ladybird is also closely linked with built-up areas.

The effect on insect populations of our notoriously unpredictable weather has perhaps been underestimated. There have been enormous changes in the population density of the common 7-spot Ladybird over the last 40 years. It was found only rarely at Kew in the years following the severe winter of 1963, but then increased to an abundance of plague proportions in 1976. Further minor fluctuations in numbers followed during the 1980's and 1990's, but the population was again reduced to a low level following unfavourable weather in the spring of 1999. It is also possible that extended periods of frost in winter may play a part in restricting the 11-spot Ladybird to milder coastal areas, but this needs further investigation.

# WHAT IS A LADYBIRD?

Everybody thinks that they know the ladybirds and certainly this is true for two of the commonest species, those familiar bright red insects with black spots that are useful to farmers and gardeners through feeding on greenfly and other aphids. Confusion arises when encountering ladybirds of different colours and shapes, or other insects with the same pattern of black spots on a red background, so it is best to start from basic principles. Nowadays, every child is taught to recognise an insect as a creature with six legs and a jointed body divided into head, thorax and abdomen.

Two special features show us that our ladybird is a beetle. On the underside of the head there is neither the curly proboscis of a butterfly nor the stout sucking beak of a bug, but two robust toothed jaws that grasp and bite the prey. But the most obvious character of beetles is the replacement of the fore-wings by hard cases that cover and protect the delicate hind-wings when the beetle is at rest. Few other insects have these hard wing-cases and those that do, such as earwigs, are unlikely to be confused with ladybirds. In most species of ladybirds, the true wings can be flicked out quickly and used for flight.

The ladybirds form a very small group among the beetles. In Britain we have only about 40 species in the ladybird family (Coccinellidae) among 4000 different species of beetles. Most ladybirds can be recognised at once from their rounded, hump-backed shape and wing-cases with a pattern of spots. Some other beetles have a similar shape or a similar pattern and colour (*Plate 16*), and some ladybirds look rather different, so it is helpful to know the special features that separate the ladybird family from other beetles. There are two of these – the shape of the antennae and the number of segments in the feet (*Figure 1*). In ladybirds, the last few segments of the antennae are swollen slightly, forming a club. If the end of the antenna is strongly swollen into a spherical club or has comb-like branches, the beetle is not a ladybird. If the antennae are slender throughout their length, this insect also is not a ladybird but probably a leaf-beetle (Chrysomelidae).

The legs of ladybirds are comparatively slender, for these are running insects – they never jump. As in all insects, each leg is divided into femur, tibia and a many-jointed foot, known as the tarsus, which ends with two claws. The tarsus is apparently formed of three distinct segments in ladybirds, but four or five segments in many other beetles. The second segment of the ladybird tarsus is large and strongly lobed. This lobe hides a minute additional segment that can only be seen with an extremely strong lens, so its presence can be ignored.

Most ladybirds can be identified from the colour and pattern of spots on the wing-cases. When in doubt, it is helpful to examine features of the thorax, both above and below. Two sections of the thorax are visible from above (*Figure 1*). The front section is expanded into a robust plate protecting the neck of the insect – this is the pronotum. The middle section of the thorax is only visible as a tiny triangle at the base of the wing-cases – this is known as the scutellum.

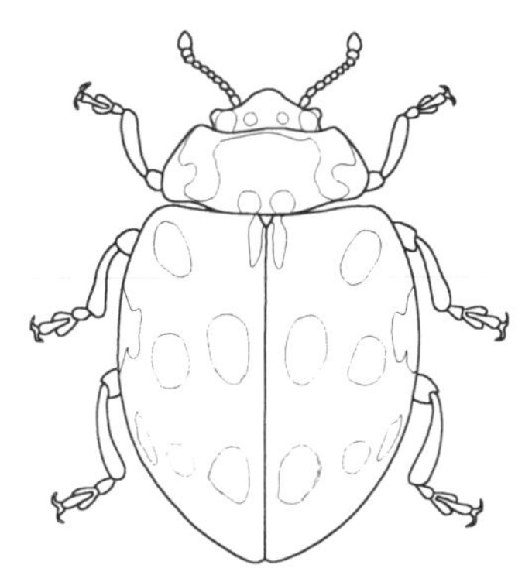

***Figure 1. Adult ladybird, from above***

For absolute certainty in identification, one must look at the underside of the insect. Between the flattened edges of the wing-cases and the true underside of the thorax, parts of the side-plates of the thorax can just be seen. In some species these offer a contrast in colour to the remainder of the underside (*Figure 2; Plates 9, 10*). Many species have a small white triangle on each side outside the base of the middle legs. Other species have, in addition to this, an even smaller white triangle in line with the base of the hind legs. Yet more species have a white bar along the sides of the thorax between the white spots.

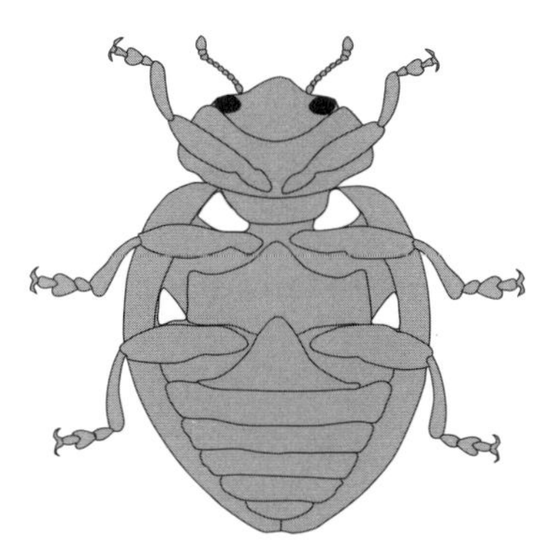

***Figure 2. Adult ladybird, from below***

Especially when examining the smaller ladybirds, two additional structures of the underside are used for separating the species. Between the bases of the front legs there may be two low ridges, usually parallel or almost so – these are the prosternal keels. On the first segment of the abdomen, behind the base of the hind legs, there are usually distinct semicircular grooves or ridges – these are the postcoxal lines. A strong lens is needed to see these minute features.

The ladybird larva is usually flat, elongate and lizard-like with outstretched legs (*Figure 3*). It must appear something like a dragon to its aphid prey. Behind the head, the first three segments of the body are broader and have legs attached – these form the thorax. The remaining segments, somewhat narrower, form the abdomen. Each segment of the abdomen has six projections that have been variously called warts, humps, lumps or bumps, but in this book are called tubercles. Each tubercle has an arrangement of spines and hairs that varies from a few short hairs to a tall branched spine. Both the colour and the spininess of the tubercles are very important for distinguishing the larvae of the different species of ladybirds.

If a short-spined larva is found that looks somewhat similar to a ladybird larva but is eating holes in leaves, then this is probably the larva of a leaf-beetle. The two ladybirds that feed on leaves have yellowish larvae with tall bristly spines (*Plate 8*).

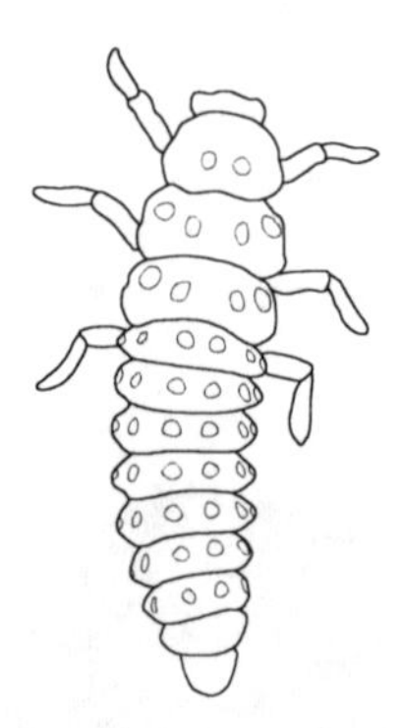

***Figure 3. Larva of ladybird***

# LADYBIRD LIFE-HISTORY

Perhaps the most remarkable thing about insects is their ability to change their shape and appearance completely as they grow up. Ladybirds, like other beetles, go through four completely separate stages during their lives: the egg; the active crawling larva; the pupa which, although apparently inert and fixed down at one end, is still capable of making sudden swift jerks in response to disturbance; and the adult ladybird which we know so well.

All creatures that live in this country and similar latitudes, not just insects, have the difficulty of getting through the winter, that colder season when the earth is tilted away from the sun and the climate tends towards the Arctic. Most insects are dormant at this time, a process that we call hibernation. All species of ladybirds have a fundamentally similar life-history, passing the winter as adults, then feeding, mating and laying eggs in the spring. The eggs hatch in about a week and the small larvae hunt for suitable food, the nature of which varies enormously between the different species. Although the larvae go through several intermediate moults (usually three), the larval stage is also not long, consisting of only a few weeks (2 to 4 weeks might be considered normal). The full-grown larva stops feeding, anchors itself at its rear end, arches its body and slowly changes into a pupa, taking several days before the skin splits and the pupa is seen. Although pale at first, the pupa darkens in colour and may have a characteristic pattern in some species. This stage also lasts for a week or a little longer before the pupal skin splits and the adult emerges. The freshly-emerged adult is pale and often unrecognisable as to species, but usually stays close to the discarded pupal skin until its new skin hardens and the colours and pattern develop. The young adult needs to feed and may move away from the pupal skin to do so, but what it does next is something of a problem. In laboratory studies, many species will mate and start the cycle of reproduction again, but this does not necessarily happen in the wild. Food may be scarce or the weather may be unsuitable for rearing another batch of larvae. The adult ladybird may hide away in shady places during warm spells in summer, a process that we call aestivation. This may be usual in warmer climates but can also occur here, especially in hot dry summers.

The timing of the reproductive cycle differs somewhat between the different species. The Pine Ladybird starts to breed in earliest spring, with mating observed frequently in March and April, although the breeding cycle takes longer than most other species and the new generation may not appear until July or August. Many species mate and lay eggs in May and June but the Orange Ladybird breeds very late in the year and no larvae have been observed before July. Most adults of this species emerge in August and September but some may not do so until October or even November. It is just possible that a very few Orange Ladybirds can get through the winter as larvae and/or pupae, although it is certain that the great bulk of its population overwinters in the adult form. These exceptionally late early stages may represent a second generation.

Some of the common ladybirds, such as the 2-spot, 7-spot and 10-spot, begin to mate quite early in the spring, typically in April, with the new generation appearing in June or July. These young adults may mate again and produce a second generation during the year, but this seems to vary between the different species and also from one year to the next. This aspect is discussed separately in the accounts of the individual species but a brief summary can be given here, based solely on observations in Surrey during the last 20 years.

There is strong evidence that the 10-spot Ladybird produces a substantial second generation in most years, since there are often two flushes of freshly-moulted adults about two months apart. Some other aphid-feeding ladybirds may produce a small, partial second generation and it is possible that the 14-spot and Adonis Ladybirds do this on a regular basis. The 2-spot has a partial second generation in some years but certainly not every year, while the evidence for a second generation in the 7-spot is minimal (during the years of this survey). As far as can be ascertained, the other ladybirds have only a single generation a year, quite certainly for the Pine and Kidney-spot Ladybirds. The 22-spot is a difficult case, since larvae have been observed from June to October, but this may just be the result of adults laying eggs over a long period.

# FINDING LADYBIRDS

Some ladybirds have bold warning colours of red and black and make no attempt at concealment. The 2-spot and 7-spot may often be encountered on a casual inspection of the garden or the neighbourhood, and perhaps also one or both of the yellow-and-black species, the 14-spot and 22-spot. Yet there are 41 species resident in Surrey, so where are all the others? In order to find the different kinds of ladybirds, two techniques are most important – searching and beating. All the larger species can be found by searching, but one must know where to look for them and this is very dependent on the time of year.

Ladybirds in winter, like dormice and overwintering butterflies, are best left undisturbed in their hibernation sites, but on the very first warm day of spring, usually in March but sometimes even in February, black ladybirds become active on tree trunks. It is important to recognise the species of tree for the ladybirds are very selective, but the Pine and Kidney-spot Ladybirds can be found most easily on the trunks of ash and sallow respectively. Examining the trunks throughout the year will reveal the full life-history of the ladybirds – feeding, mating, egg-laying, development of larvae, pupation and emergence. The 2-spot and 10-spot Ladybirds also occur on the trunks of broad-leaved trees in spring but generally disappear among the foliage as soon as the leaves come out. The 7-spot may often be found at the foot of the trees but it generally stays close to the ground.

Searching among foliage is particularly rewarding in June and July, because ladybirds may be found at different stages of development. Well-grown larvae, pupae and adults, sometimes of two overlapping generations, may all be present at the same time. Careful searching leaves the insects undisturbed and allows the observation of natural behaviour such as feeding and pupating. The foliage of trees, tall herbaceous plants and grasses may all be examined with good prospects of finding ladybirds, particularly if aphids are present. Aphid colonies are often revealed by a trail of ants intent on milking them for honeydew.

Since pine trees are evergreen, ladybirds may be found among the shoots and needles at any time of year. Dormant ladybirds often cluster around the ends of shoots or hide beneath the stalks of young cones. There may be six or more different species on a Scots pine in a park or garden anywhere in Surrey, and up to ten species on pines on the heathy commons of west Surrey.

Many ladybirds are small, so one needs to get close to them and, for this reason, children are particularly good at spotting ladybirds. Adults perhaps just need to sit down and relax – find a grassy bank, anywhere in the countryside on a warm day, and one will see not just ladybirds but many kinds of insects and other creatures going about their business. The ladybirds are likely to include adults or larvae of the 16-spot and 24-spot Ladybirds and, if one is extremely lucky, the larvae of various species of *Scymnus* that resemble small white hedgehogs. It may be necessary to convince one's companions that insect-watching is a serious occupation and not just an excuse for laziness.

The similar larvae of *Scymnus auritus* may occasionally be found beneath the leaves of oak trees in July. These leaves and many others become covered in white powdery mildew from July onwards. Both adults and larvae of the 22-spot Ladybird feed on this mildew and are most easily found on hogweed leaves in August. From mid-August onwards, turning over the mildewed leaves of sycamore and other broad-leaved trees will often reveal larvae, pupae or freshly-emerged adults of the Orange Ladybird.

Beating is an entomologist's short-cut to making a quick record of the species in a particular habitat at a given site. The beating tray is a white sheet stretched over a wooden frame and held under the branches of a tree or shrub, which is then tapped firmly with a robust stick. The insects on the branch fall onto the sheet and can be examined there. Afterwards, the insects can be returned to their habitat by inverting the beating tray and tapping the cloth gently with the stick.

Beating is a particularly valuable technique in autumn and in spring, especially in cool dry weather when the ladybirds are on the plants but not very active, often sheltering beneath leaves. The most reliable method of recording the 10-spot and Cream-spot Ladybirds is by beating hawthorn in spring, either just before or just after the unfolding of the leaves. In order to find all the species associated with pine, it is sometimes necessary to beat many trees, but beating is often the only way to find the smaller ladybirds that might otherwise be overlooked.

Beating ladybirds from ivy needs a technique all of its own, since the woody climbing stems lie closely against the supporting tree. The stick is held horizontally and used to brush the leaves downwards. Many species of ladybirds can be found sheltering among ivy leaves.

The beating tray is not only useful for surveying the fauna of trees and shrubs. It can be placed beside a clump of nettles, thistles or other tall plants which are then given a very gentle sideways tap. This method can even be used for grass tufts. The tray is placed at an angle with the sheet inserted as far beneath the tuft as it will go. The 14-spot, 16-spot, 22-spot, 24-spot and several smaller ladybirds can all be recorded this way, sometimes in large numbers.

The entomologist's other favourite technique is sweeping a sturdy net from side to side while walking slowly through grass or other vegetation. All the species just mentioned can be found by sweeping, with the disadvantage of not knowing exactly from which plant they have come.

Various methods for trapping insects are used in entomology. All of them catch a few ladybirds, but none are as effective and satisfying as recording the ladybirds in their natural habitat. Nevertheless, the traps often take species that may otherwise be difficult to find so it will always be worth-while to monitor the catch. The Orange Ladybird is a regular visitor to the mercury-vapour lights used for recording moths, so if your local moth recorder has not seen this species, then it is undoubtedly absent from the neighbourhood. Several conifer specialists including the 18-spot also appear at mercury-vapour light (GAC) and the small species of the genus *Scymnus* occasionally turn up at the low-intensity light-traps of the Rothamsted network (AJH). The fixed Malaise traps catch insects that fly upwards and over an obstacle, mainly flies and parasitic wasps but also a few ladybirds including the smaller species. Flight-interception traps catch insects such as beetles that drop down on encountering an obstacle, and so catch many ladybirds if operating at their time of dispersal. Pitfall traps catch insects that walk over the ground and, surprisingly, caught large numbers of larvae of arboreal ladybirds in a spruce forest in Germany (Klausnitzer & Bellmann, 1969).

## IDENTIFICATION OF LADYBIRDS

The identification of ladybirds is easy, and most species can be named on sight. It is initially a matter of counting the spots, or matching the pattern to the photographs. The number and arrangement of spots is fairly constant in most species. There are just three species, normally red or orange with black spots, that have melanic varieties that are black with red spots as well as chequered varieties with equal amounts of red and black. These are the 2-spot, 10-spot and Hieroglyphic Ladybirds. One learns to recognise these melanic and chequered varieties as if they were different ladybirds. A few other species show some variation in pattern, yet usually remain recognisable.

The difficulties caused by variation have been greatly exaggerated in the past, usually by experts working in museums. The collections of insects in natural history museums include all the unusual ladybirds found by beetle collectors over the last 200 years and may give a false impression of the frequency of such varieties in the wild. Also, ladybirds are often encountered in numbers, so an unusual variety may be feeding alongside or even mating with a typical specimen.

One small but important point is that an unidentified ladybird should be kept as a temporary captive in a small, preferably transparent container such as a plastic tube or box. Otherwise it is likely to fly away. The insect can be examined inside the container using a 4x or 10x lens kept in a pocket or around one's neck, but younger readers may be able to see all the significant features with the naked eye.

The reader of this book is encouraged to name all 24 species of the larger ladybirds accurately. Identifications made from the colour plates can be confirmed or rejected by reading the first two or three paragraphs about the species, which deal with the extent of variation, additional features to check, and any possible confusion with other species. It is always helpful to read all about a creature just identified. This systematic approach, of becoming thoroughly familiar with one species at a time, is useful to anyone learning about the natural world and can be applied to animals, birds, wild flowers and other insects.

The smaller ladybirds need specialised entomological techniques for their identification. It is usually necessary to collect a specimen, examine it under a binocular microscope and, for some species, dissect out its internal parts. Specimens can be named using the main key to adults in Majerus & Kearns (1989). However, even with these small insects, the likely identification can often be guessed from its size, colour and habitat.

The identification of ladybird larvae is largely based on the arrangement of spines and hairs on the body, which is different for each species but usually needs to be examined through a lens. Many of the species also have a unique pattern of coloured spots. The larva is completely different from the adult in the colour, number and grouping of its spots, with the single exception of the 22-spot Ladybird. About 15 of the larger ladybirds have distinctive larvae which can be named using this book, but all can be identified through the key to larvae in Majerus & Kearns (1989) which uses structural details as well as colour pattern. The larvae of most of the smaller ladybirds can be named using the more comprehensive key by van Emden (1949), or that in Hodek (1973) which also includes many non-British ladybirds.

The identification of ladybird pupae has not yet been worked out in detail. Many of them appear to have distinctive colour patterns and structural features such as a covering of hairs, or teeth on the side of the abdominal segments. There is a basic division into those pupae which lie within the split larval skin and those which protrude from the crumpled larval skin, either wholly or partly. A photographic study of the pupae of all species might well reveal sufficient differences to enable their identification.

# IDENTIFICATION STARTERS

Most ladybirds may be readily identified, initially from the photographs and then by checking the text. Occasional specimens may not match any of the ladybirds illustrated on the colour plates. In such cases an alternative starting point is offered through the following lists – the species are grouped according to the pattern on the wing-cases, the number of spots on the underside of the thorax, other distinctive features, size and habitat. It should be noted that immature specimens are paler – the red colour may remain orangey for many months and even the black colour is brown at first. The smaller ladybirds are only included in the lists when they are closely associated with a particular habitat.

## 1. PATTERN OF WING-CASES

***Red with black spots*:** **2-spot** (p.65); **5-spot** (p.78); **7-spot** (p.72); **Scarce 7-spot** (p.75); **11-spot** (p.79); **Adonis** – usually 7 to 11 spots (p.56); **Eyed** – usually 18 spots with pale rims (p.90); **Water** – 19 spots (p.60); **24-spot** – with short hairs (p.43).

***Orange-red with black spots*:** Immature specimens of any of the above; **10-spot** (p.69); **Bryony** – with short hairs and 11 spots (p.41); **Cream-streaked** – either 4 or 16-20 spots and background with pale streaks (p.81); **Hieroglyphic** – elongate spots (p.77).

***Orange-red without spots*:** **Larch** (p.59); **10-spot** (p.69); **24-spot** (rare variety) – with short hairs (p.43).

***Orange-brown with whitish spots*:** **Orange** – front of thorax straight (p.92); **Cream-spot** – front of thorax indented (p.84); **18-spot** – basal spots L-shaped (p.82); **Striped** – elongate spots (p.89).

***Yellow with black spots*:** **22-spot** (p.96); **14-spot** – inner spots usually linked (p.86).

***Cream with black spots*:** **16-spot** – outer spots linked (p.62); **14-spot** (p.86); **Water** (immature) (p.60).

***Black with red spots*:** **Pine** – 4 spots with front pair comma-shaped (p.51); **Heather** – spots in narrow transverse row (p.46); **Kidney-spot** – 2 large spots (p.48); [all with wing-cases turned outwards at edges] **2-spot** – usually 4 or 6 spots (p.65); **10-spot** – comma-shaped spots on shoulders (p.69); **Hieroglyphic** – few or no spots (p.77).

***Chequered pattern*:** **10-spot** (p.69); **14-spot** (p.86); **2-spot** (p.65); **Hieroglyphic** (p.77).

## 2. UNDERSIDE OF THORAX AND ABDOMEN

***Entirely black*: 2-spot** (p.65); **Hieroglyphic** (p.77).

***Black with red tip to abdomen*: Pine** (p.51); **Heather** – red tip and sides (p.46); **Kidney-spot** – most of abdomen red (p.48). [all with wing cases turned outwards at edges]

***Entirely reddish-brown*: 24-spot** (p.43).

***Black or brown with one pair of white or yellowish spots outside base of middle legs*: 5-spot** (p.78); **7-spot** (p.72); **10-spot** (p.69); **11-spot** (p.79); **Cream-spot** (p.84); **Striped** (p.89); **Eyed** – with trace of second pair of spots (p.90).

***Black or brown with two pairs of white or yellowish spots outside middle and hind legs (Figure 2 and Plate 9)*: Scarce 7-spot** (p.75); **Adonis** (p.56); **Water** – legs pale (p.60); **14-spot** – second pair small (p.86).

***Brown with two pairs of white or yellowish spots linked by pale side bar (Plate 10)*: Cream-streaked** (p.81); **18-spot** (p.82); **22-spot** – side bar partially pale (p.96); **16-spot** – side bar partially pale (p.62); **Larch** – side bar partially pale (p.59); **Orange** – spots and side bar slightly paler than ground colour (p.92); **Bryony** – sides of thorax and abdomen entirely orange-red (p.41).

## 3. OTHER FEATURES

***Wing-cases turned outwards at edges*: Pine** (p.51); **Heather** (p.46); **Kidney-spot** (p.48); **Orange** (p.92).

***Covered with short hairs*: 24-spot** (p.43); **Bryony** (p.41); All the smaller ladybirds except *Hyperaspis*.

***Thorax broadest in middle*: Adonis** (p.56); **Water** – legs pale (p.60).

## 4. SIZE (LARGEST FIRST)

***Large (7-8 mm)*: Eyed** (p.90); **Striped** (p.89); **7-spot** (p.72); **Scarce 7-spot** (p.75); **Bryony** (p.41); **Cream-streaked** (p.81).

***Medium large (5-6 mm)*: Orange** (p.92); **Cream-spot** (p.84); **2-spot** (p.65); **5-spot** (p.78).

***Medium (4-5 mm)*: Kidney-spot** (p.48); **Hieroglyphic** (p.77); **Pine** (p.51); **14-spot** (p.86); **Larch** (p.59); **18-spot** (p.82); **10-spot** (p.69); **11-spot** (p.79); **Adonis** (p.56); **Water** (p.60).

***Small (3-4 mm)*:** **22-spot** (p.96); **24-spot** (p.43); **Heather** (p.46); **16-spot** (p.62).

***Very small (1-3 mm)*:** All other species (smaller ladybirds).

## 5. HABITAT

***Pine trees*:** **Eyed** (p.90); **Striped** (p.89); **Cream-streaked** (p.81); **Pine** (p.51); **Larch** (p.59); **18-spot** (p.82); **10-spot** (p.69); ***Scymnus suturalis*** (p.113); ***Scymnus nigrinus*** (p.108).

***Other conifers*:** **Pine** (p.51); **Larch** (p.59); **18-spot** (p.82); **10-spot** (p.69).

***Foliage of broad-leaved trees*:** **Cream-spot** (p.84); **10-spot** (p.69); **2-spot** (p.65); **Orange** (p.92); ***Scymnus auritus*** – oaks (p.110).

***Trunks of broad-leaved trees*:** **Pine** (p.51); **Kidney-spot** (p.48); **2-spot** (p.65).

***Tall herbaceous plants*:** **14-spot** (p.86); **22-spot** (p.96); **2-spot** (p.65); **7-spot** (p.72); **11-spot** (p.79); **Adonis** (p.56).

***Grassland*:** **22-spot** (p.96); **24-spot** (p.43); **16-spot** (p.62); **7-spot** (p.72); ***Rhyzobius litura*** (p.101).

***Wet places*:** **Water** (p.60); ***Coccidula rufa*** (p.99); ***Coccidula scutellata*** (p.100).

***River shingle*:** **5-spot** (p. 78).

***Near nests of wood ants*:** **Scarce 7-spot** (p.75).

***Heather*:** **Hieroglyphic** (p.77); **Heather** (p.46).

***Leaves of white bryony*:** **Bryony** (p.41).

***Ivy*:** ***Nephus quadrimaculatus*** (p.114); ***Clitostethus arcuatus*** (p.117); Overwintering examples of many other species.

***Feeding on mildew*:** **22-spot** (p.96); **Orange** (p.92); **16-spot** – also on pollen (p.62).

***Feeding on leaves*:** **24-spot** (p.43); **Bryony** (p.41).

Fully illustrated keys for the identification of all the larger ladybirds are given by Moon (1986) and by Majerus & Kearns (1989).

# LADYBIRDS AS PETS

Keeping insects as pets has little in common with looking after dogs, cats or horses, but is more like keeping birds in a cage or goldfish in a bowl. It has, however, the more serious purpose of finding out something about the insect. This may merely be a matter of rearing a larva to the adult stage to establish exactly which species it is. The excitement of making a personal discovery should not be underestimated, even if others have done it before. After 16 years I still remember the fascination of finding two distinct types of ladybird larvae on the trunks of ash trees, but not knowing how they corresponded to the two species observed there previously as adults. Then in 1986 came the shock of finding that an Orange Ladybird had emerged from a pupa that I had not even recognised as a ladybird.

The pupa on a leaf is by far the simplest type of ladybird to rear, since it requires no attention and should produce an adult within a week or two. Once reared and identified, adult ladybirds can be released, if possible at their place of capture. Failing this, I make a point of only releasing specimens into a known population (see distribution maps) and, above all, one should never release an insect in this country if it was captured overseas.

Small insects being reared can be kept individually in small plastic boxes (mine are 8 cm x 5 cm x 2 cm). They are best kept away from direct sunlight in an unheated room. A twist of damp tissue helps to keep foliage fresh and provides moisture for the insect.

Larvae are more difficult to rear since they need food. The larger they are when captured, the less food they will need. Fresh food needs to be supplied every few days. This is best done by transferring the larva (with any associated labelling) to a clean box containing the new supply of food. The technique can also be used to rear other kinds of insect, although each type has its own particular problems.

The mildew-feeding ladybirds are easy to rear, as are those that feed on the leaves of plants. The aphid-feeders need a regular supply of prey, which need not be from their original host-plant. Some trial and error is often necessary, since some aphids may be acceptable and others not, and the menu varies between the different species of ladybirds. For those ladybirds that feed on scale-insects, some ingenuity is needed to find them suitable food.

The exact food of several species is not known so it is quite possible that an amateur entomologist, experimenting with a larva in a little box, may make a significant discovery.

Much more information about keeping ladybirds in captivity is given by Majerus & Kearns (1989).

# THE NAMES OF LADYBIRDS

Only a few insects have English names but these include all the larger ladybirds. Some of these names are very old, such as 7-spot Ladybird, while others were invented for the Cambridge Ladybird Survey and one, the Bryony Ladybird, is used here for the first time.

Many names just refer to the number of spots. Where two or more species have the same number of spots, other names are needed. These may be descriptive of the colour or the markings, as in the Orange, Striped, Eyed, Hieroglyphic, Cream-streaked, Cream-spot and Kidney-spot Ladybirds, or refer to the habitat or the host-plant, as in the Water, Larch, Pine, Heather and Bryony Ladybirds. The names derived from host-plants can be a little confusing. The host of the Bryony Ladybird is the white bryony, not the totally unrelated black bryony, while the other names refer to the principal host, although the ladybirds can also feed elsewhere. Thus the Larch Ladybird occurs not only on larch, but also on spruce, Douglas fir and pine. The Heather Ladybird is occasionally found on other hosts, principally juniper in Surrey. The most confusing name is that of the Pine Ladybird, now one of our commonest species. If one goes back a hundred or even fifty years, it was known almost exclusively from pine and perhaps a few other conifers, but since then it has spread to broad-leaved trees, in particular to the trunks of ash. More recently still, it has been found in increasing numbers on the trunks of horse-chestnut, sycamore and other trees in association with an introduced species of scale-insect. It is thus an everyday occurrence to find Pine Ladybirds among horse-chestnut scales on a sycamore trunk.

One exception to the pattern of English names is the Adonis Ladybird. This name was given to match the scientific name of its genus, *Adonia*. Unfortunately, a change in scientific opinion has led to the ladybird being placed in a different genus, *Hippodamia*, but such changes should not be allowed to affect the English names.

Scientific names are written in a language unfamiliar to most people, being classical Latin with some word-elements taken from ancient Greek. This acts as a deterrent to many people who are put off by the difficulty of understanding and pronouncing them. Yet the same people will happily use the scientific names of garden flowers such as *Fuchsia* and *Rhododendron*, or of prehistoric monsters such as *Tyrannnosaurus*, *Triceratops* and other dinosaurs. It is all a question of familiarity.

The scientific names of ladybirds are easier to understand than those of most other creatures, for many of them are direct translations of the English names, although in almost every case it was the scientific name that came first! Thus the 7-spot is *Coccinella 7-punctata* and the Hieroglyphic Ladybird is *Coccinella hieroglyphica*, and so on.

The first word of the scientific name refers to the genus, or group of closely-related species, while the second word is the specific or individual name of the species. All ladybirds were originally placed in the genus *Coccinella*, so all the specific names had to be different. The old genus *Coccinella* has now been split into many smaller genera such as *Adalia*, *Anatis* and *Calvia*, and only four of our species remain in *Coccinella*.

At first ladybirds were given numeric names such as *bipunctata* and *7-punctata*, but even after using almost all numbers up to 24, it was clear that there were far more different species of ladybird than there were available numbers, especially since the world-wide fauna was being considered. This problem was partly overcome by using descriptive names, some of which are direct translations of the English names such as *ocellata* for the Eyed Ladybird and *renipustulatus* for the Kidney-spot Ladybird. Another method used to create more names was to replace *punctata*, meaning spotted, with other words of similar meaning, as we might use dotted, speckled or even blotched. These alternatives are *maculata*, *guttata*, *notata* and *pustulata*. Thus we have *bipunctata* for the 2-spot but *bipustulatus* for the Heather Ladybird. The difference in ending is a feature of Latin and most modern European languages, the agreement in gender between an adjective and its noun. In its simplest form this is merely a matter of rhyming, *-a -a* (feminine) or *-us -us* (masculine), giving *Adalia bipunctata* and *Chilocorus bipustulatus* as the names of these two species.

The scientists who created the system of names, and continue to refine it, were and are exceedingly clever people, so it is remarkable that they have invented a system in which the official name of almost every species is changed every few years, like a chamaeleon changing its colour. It is this constant changing of names, rather than any difficulty of understanding or pronunciation, that is the great drawback of the system. This is not a particularly big problem with the ladybirds. There are two kinds of change: species that alternate between two different generic names, and specific names subject to sequential change as first one name, then its replacement, is found to be incorrect. The first type is represented by *22-punctata* switching between *Thea* and *Psyllobora*, and *16-punctata* alternating between *Micraspis* and *Tytthaspis*. The Scarce 7-spot belongs to the second type, with *divaricata* replaced by *distincta* and then by *magnifica*. The most important of these alternative names (historical synonyms) are given at the top of each species account.

The scientific names are written in Latin and go back to the foundation of the system by Linnaeus, a Swede who spoke and wrote in Latin, as did all scientists and educated men of his day. The use of Latin as a language of international communication is not as dead as might be supposed, since almost all the technical and scientific words used in biology and medicine have their origin in Latin and Greek. It is unfortunate that this is not realised by our educators. The introduction of courses in basic Latin and Greek for all biology students is long overdue.

A complication peculiar to ladybirds arises because the scientific names are meant to be used by people speaking different languages. We would naturally read out *7-punctata* as *seven-punctata* while a Frenchman would say *sept-punctata* and a German *sieben-punctata*. To overcome this difficulty it has been decreed that the names of the ladybirds should be written out in full in their original language, Latin. So the official name of the 7-spot Ladybird is *Coccinella septempunctata*. In some cases this makes the name very long and clumsy, and few people nowadays are able to count up to 24 in Latin, so the abbreviated names such as *7-punctata* are used throughout this book. The full Latin names are given below.

| | | |
|---|---|---|
| *Coccinella quinquepunctata* | *C. 5-punctata* | 5-spot |
| *Coccinella septempunctata* | *C. 7-punctata* | 7-spot |
| *Adalia decempunctata* | *A. 10-punctata* | 10-spot |
| *Coccinella undecimpunctata* | *C. 11-punctata* | 11-spot |
| *Hippodamia tredecimpunctata* | *H. 13-punctata* | 13-spot |
| *Propylea quatuordecimpunctata* | *P. 14-punctata* | 14-spot |
| *Calvia quatuordecimguttata* | *C. 14-guttata* | Cream-spot |
| *Tytthaspis sedecimpunctata* | *T. 16-punctata* | 16-spot |
| *Halyzia sedecimguttata* | *H. 16-guttata* | Orange |
| *Myrrha octodecimguttata* | *M. 18-guttata* | 18-spot |
| *Anisosticta novemdecimpunctata* | *A. 19-punctata* | Water |
| *Psyllobora vigintiduopunctata* | *P. 22-punctata* | 22-spot |
| *Subcoccinella vigintiquatuorpunctata* | *S. 24-punctata* | 24-spot |

***Full Latin names of ladybirds commonly abbreviated***

# THE ABUNDANCE OF LADYBIRDS

Which is the commonest species? Someone starting to learn about ladybirds will generally first come to recognise the 2-spot, since it enters houses and is most frequent in towns, then the 7-spot from occurring in gardens and on roadsides, and probably next the 14-spot with its strikingly different colour pattern of yellow and black. It will come as a surprise to find other, less familiar species at the top of our list of most abundant ladybirds. These live in habitats less often examined, such as in grassland close to the ground, in stands of long grass and on tree trunks.

The following table ranks the species according to the total number of specimens reported during the 20 years of the survey. This method will tend to underrate the common species because many records refer to their presence at a particular site or in a specified tetrad and give no indication of numbers. Such records have been included as one specimen although many may have been seen. The figures do however give a rough idea of the abundance of the different species.

| | | | |
|---|---|---|---|
| 1. | *Tytthaspis 16-punctata* | 16-spot Ladybird | 28,500 |
| 2. | *Exochomus quadripustulatus* | Pine Ladybird | 5,450 |
| 3. | *Coccinella 7-punctata* | 7-spot Ladybird | 4,980 |
| 4. | *Adalia bipunctata* | 2-spot Ladybird | 3,715 |
| 5. | *Subcoccinella 24-punctata* | 24-spot Ladybird | 2,769 |
| 6. | *Propylea 14-punctata* | 14-spot Ladybird | 1,370 |
| 7. | *Psyllobora 22-punctata* | 22-spot Ladybird | 1,222 |
| 8. | *Chilocorus renipustulatus* | Kidney-spot Ladybird | 1,183 |
| 9. | *Rhyzobius litura* | | 918 |
| 10. | *Adalia 10-punctata* | 10-spot Ladybird | 842 |
| 11. | *Halyzia 16-guttata* | Orange Ladybird | 470 |
| 12. | *Calvia 14-guttata* | Cream-spot Ladybird | 369 |
| 13. | *Harmonia quadripunctata* | Cream-streaked Ladybird | 318 |
| 14. | *Anisosticta 19-punctata* | Water Ladybird | 227 |
| 15. | *Anatis ocellata* | Eyed Ladybird | 186 |
| 16. | *Scymnus suturalis* | | 127 |
| 17. | *Myrrha 18-guttata* | 18-spot Ladybird | 126 |
| 18. | *Aphidecta obliterata* | Larch Ladybird | 125 |
| 19. | *Coccidula rufa* | | 124 |
| 20. | *Chilocorus bipustulatus* | Heather Ladybird | 90 |

***The twenty most abundant ladybirds in Surrey***

It is interesting to compare this table with that derived from the national distribution as recorded by the Cambridge Ladybird Survey (Majerus *et al.*, 1997). The 16-spot rises from 13th to 1st place through counting specimens rather than the number of 10-km squares where it occurs; the strongly south-eastern bias to its distribution gives it a low place nationally. The Pine Ladybird rises from 11th to 2nd place, again through counting specimens, although I suspect that some Cambridge recorders did not look for it on broad-leaved trees. The small *Rhyzobius litura* rises from 22nd to 9th place, but only because many Cambridge contributors did not record this small beetle while a special search was made for it in Surrey. The Orange Ladybird rises from 18th to 11th place, almost certainly because it is genuinely increasing in range and numbers. The 11-spot falls from 7th on the national list to about 25th in Surrey, chiefly through being a common species in coastal areas but infrequently found inland; counting 10-km squares may also overestimate its abundance nationally, as it certainly does in Surrey.

## CONSERVATION

It is a pleasure to be writing about a group of insects that are doing well in Britain. Most of the ladybirds are extremely mobile and highly adaptable with an ability to attack new pests. They can often find small patches of suitable habitat and survive there.

Our list for the vice-county can be compared directly with one from nearly a hundred years ago published in the *Victoria History of the County of Surrey* (Goss, 1902). Since that date it appears that we have only lost one species, the 13-spot Ladybird which may never have been a permanent resident but only an immigrant forming temporary colonies in favourable years (Majerus, 1994). One small species, *Hyperaspis pseudopustulata*, was last found in 1975 just before the start of our survey and is unlikely to have been lost. There have, on the other hand, been several gains. The Cream-streaked Ladybird arrived in the 1950's and three other species have established themselves in the 1990's. The tiny *Nephus quadrimaculatus* has long been known from elsewhere in Britain but the Bryony Ladybird and *Rhyzobius chrysomeloides* are new arrivals from the Continent.

Our most threatened habitats are chalk downland, heathland and marsh. Two minute black species, *Scymnus schmidti* and *S. femoralis*, are restricted to dry places on chalk and sand. They are certainly still present but appear to have decreased. *Platynaspis luteorubra* is another small species for which our only two records come from chalky soil. These species are found only rarely, as are some other small ladybirds, but they are so elusive and so little is known about them that no recommendations for their conservation can be made beyond maintaining the habitat in which they occur.

Two species are exclusive to heathland, the Heather and Hieroglyphic Ladybirds. I believe that the latter is our most threatened species and its food preferences need further investigation. Although the loss of heathland is caused largely by invading pine trees, it is ironic that these same pines form one of our most important habitats for ladybirds, as well as many other insects.

The Scarce 7-spot Ladybird is found almost exclusively on partly wooded heath where it is one of a large community of insects, principally beetles, that live alongside the wood ant, *Formica rufa*. This community needs to be conserved as a whole. The wood ant appears to be favoured by coppice management, i.e. cutting back the trees at regular intervals.

The extinct 13-spot was predominantly a wetland species but there are three resident ladybirds that favour waterside vegetation and marshy land. The last 20 years have seen several periods of drought when these wetland species became difficult to find. More normal weather conditions at the end of the survey have seen them become reasonably common again.

The 11-spot and Adonis Ladybirds are comparative rarities in Surrey but are common insects elsewhere, the 11-spot in other parts of Britain and the Adonis in southern Europe. They are species on the edge of their range and the reasons for their scarcity here are likely to involve climate and weather. No conservation measures are necessary.

The Orange Ladybird was formerly considered something of a rarity. It is now known to feed primarily on the mildew affecting sycamore leaves in late summer and autumn. These facts should not, however, tempt the managers of nature reserves to relax their efforts to eradicate sycamore from their land. This ladybird has increased in numbers in recent years to become rather common over most of the county and is recorded as feeding on the mildewed leaves of at least five other species of tree and shrub in Surrey. Sycamore remains an invasive alien with many fewer associated insects than the native oak.

Our rarest species appears to be the tiny *Clitostethus arcuatus*, found regularly at only one site in Surrey and turning up occasionally in other parts of Britain. It is known to feed on whitefly, but the reasons for its apparent rarity need investigating.

# PEST CONTROL

Many of our commonest ladybirds feed exclusively on aphids and so occupy the front rank of natural pest-controllers along with hoverflies, lacewings and the specialised parasites of aphids.

Since most field crops are sprayed regularly with insecticides, our native ladybirds are only able to maintain their populations in marginal areas such as field edges, roadsides and waste ground, where they feed on aphids on undervalued weeds such as nettles, thistles and umbellifers. These weeds hold a reservoir of ladybirds that will spread out and attack pests when required.

The leaves of trees are largely unaffected by human activities and provide a wholly natural habitat for plant-feeding insects with their predators and parasites. The specialised ladybirds of coniferous trees in particular must be of great benefit to forestry. The 2-spot Ladybird is an important species for gardeners. It is generally abundant in towns, occurring mainly on trees but also on shrubs and herbaceous plants. It readily attacks aphids on garden plants such as roses, but many gardeners prefer to spray their roses rather than risk damage through aphids by taking a chance on biological control.

A great deal of research has been done on the effect of ladybirds in controlling aphids, but the subject is so vast and complicated that there must still be some aspects left unexplored. Aphids have many defence mechanisms and the individual species of ladybird may be differently equipped to deal with them.

One experiment in particular is going on in front of our eyes. The large scale-insect *Pulvinaria regalis* lays its eggs in unsightly masses of white waxy wool on the trunks of certain broad-leaved trees, particularly those growing in towns and on roadsides. The Pine Ladybird has now been feeding on these egg-masses for at least 15 years and appears to eat both the eggs and their white covering. But is this an effective control? The scale-insect still seems to be spreading, with new colonies being found in further towns and cities, and also on additional trees in areas where it has long been established.

# FURTHER READING

## BOOKS ABOUT LADYBIRDS

**Moon, A., 1986.**

*Ladybirds in Dorset*, a guide to their identification and natural history. 24pp, 7 black & white plates. Dorset Environmental Records Centre, Dorchester.

A simple introduction to ladybirds with a key to all the larger species and three of the smaller ones, all of which are illustrated by large detailed drawings. Available from DERC, Colliton House Annexe, Glyde Path Road, Dorchester, Dorset DT1 1XJ. Price £2.40 plus £1 postage & packing.

**Majerus, M.E.N., & Kearns, P.W.E., 1989.**

*Ladybirds*. Naturalists' Handbooks, 10. 104 pp, 4 colour & 4 black & white plates. Richmond Publishing, Slough.

The only book in print with an identification key to all ladybirds including the smaller species. Also has a simple field key to the adults and a key to the larvae of all the typical ladybirds. Much information on the biology of ladybirds.

**Majerus, M.E.N., 1994.**

*Ladybirds*. The New Naturalist series, 81. 368 pp, 16 colour plates, 184 text figures (incl. photos.). Harper Collins, London.

Large volume packed full of information about ladybirds. The 16 pages of colour plates include several species not illustrated in this book and additional aspects of variation and behaviour.

**Rotheray, G.E., 1989.**

*Aphid Predators*. Naturalists' Handbooks, 11. 78pp, 2 colour plates. Richmond Publishing, Slough.

Essential introduction for anyone wishing to investigate the relationship between aphids and their predators, including ladybirds. Includes notes on aphid species, methods of rearing and ideas for projects.

## PUBLISHED DISTRIBUTION MAPS FOR OTHER COUNTIES

**Surry, R., 1993.**
The distribution of ladybirds in Dorset. *Recording Dorset* **3** (July 1993): 2-9.
This issue available from Dorset Environmental Records Centre (address opposite), price £2.

**Campbell, J. M., 1985.**
*An atlas of Oxfordshire ladybirds.* 28pp. Oxfordshire Museums, Occasional Paper No. 8.
Available from Oxfordshire County Council, Biological Records Centre, Witney Road, Standlake, Oxfordshire OX8 7QG. Now out of date, but Oxfordshire BRC holds several thousand ladybird records on a computer database.

We offer our apologies if any local atlas of ladybirds has been overlooked.

# ACKNOWLEDGEMENTS

The series of Surrey Wildlife Atlases is a cooperative venture in which everybody helps each other out in the best amateur tradition. Roles are continually interchanging with a recorder for one book becoming editor of the next and author of a third. For *Ladybirds of Surrey* I owe a huge debt to Graham Collins who has acted as computer expert, photographer, editor, proof-reader and artist as well as being one of the chief recorders.

It is an enormous task to record the distribution of even a small family of insects on a grid of over 500 squares, but my task has been made easier since both Graham and Roger Morris, while recording for their own books in the series, have regularly produced lists of ladybirds from the sites they visited.

Many other recorders have contributed to this survey and their names are listed overleaf, but some deserve special mention. Jane McLauchlin and John Steer encouraged me to start the survey and supplied many records in its early years. Derek Coleman took on the recording of a complete 10-km square while also being its bird recorder. Andrew Halstead and Andrew Salisbury combined to ensure that the tetrad including Wisley Common and the Royal Horticultural Society's Garden has by far the highest number of ladybird species of any square in Surrey. In the last years of the survey, Jonty Denton provided a flood of records from the west of the county, particularly of the wetland species.

I am particularly grateful to those beetle specialists resident in the county who have taken the trouble to send me their records of ladybirds, although this family forms only a tiny part of the British beetles. Max Barclay, Ian Menzies and Roger Booth have all supplied many valuable records, particularly of the more difficult, smaller species. Ian has kept me in touch with his latest discoveries and Roger, for whom ladybirds are a speciality, has advised on the scientific names of the species, assisted with some difficult identifications, helped me find elusive references and encouraged the removal of many inaccuracies and ambiguities from my text. Further problems with the entomological literature have been resolved by John Muggleton, Robert Pope, Peter Chandler and John Badmin.

John Pontin and David Baldock have allowed me to publish the provisional distribution map for the wood ant, *Formica rufa*, which is being prepared for a future volume in the series of Surrey Wildlife Atlases.

As authors of previous books in this series, Graham Collins, Peter Follett and Roger Morris have allowed me to reproduce maps originally prepared for their volumes. These maps, and all the other distribution maps in this book, were prepared using the DMap computer program written by Dr Alan Morton of Imperial College at Silwood Park.

Mark Telfer of the national Biological Records Centre has allowed access to the records held on ladybirds, in particular to those of the genera *Scymnus* and *Nephus* from all the national and some private collections that were used to create the distribution maps in Pope (1973); these are published with permission from Robert Pope.

Giving lectures to local groups must sometimes be a thankless task, either preaching to the converted or trying to arouse a passive audience, so it is a pleasure to put on record that the seed of this book was sown at a well-attended and enthusiastic meeting of the East Grinstead Natural History Society on 21st April 1980. The knowledge and photographic skills of Robert Pope were put across so ably by his wife Joyce that I was filled with amazement at the colour and variety of British ladybirds and fired with enthusiasm to make a special study of this group of insects.

It is the humble duty of an author to remind his audience that every picture is worth a thousand words, and I offer grateful thanks to the photographers who have donated their superb colour slides towards this book. They are David Element, who has also acted as photographic editor, Graham Collins, Andy Callow, Richard Jones, Tony Mundell, Jim Porter, Michael Majerus and Robert Pope. My only regret is that an abundance of material and some duplication of subject have forced us to reject some absolutely outstanding photographs and I hope that these can be published elsewhere.

Michael Majerus and the Cambridge Ladybird Survey have made an enormous indirect contribution through training and enthusing many of our recorders. They also lent us the specimens shown in some of our photographs, particularly those of larvae.

Paul Wickham and a substantially new team at the Surrey Wildlife Trust have maintained the efficiency of their predecessors. Clare Windsor has now retired from the Trust but continues to work on the Wildlife Atlas Project and has guided this book through its design and production phase with her customary patience, common-sense and skill.

We gratefully acknowledge the generous sponsorship of the Surrey Biodiversity Partnership which has enabled this book to be published.

# LIST OF RECORDERS

K.N.A. Alexander (KNAA)
A.A. Allen (AAA)
D.W. Baldock (DWB)
A.J. Baldwin
M.V.L. Barclay (MVLB)
R.G. Booth (RGB)
M.D. Bridge (MDB)
J.P. Brock (JPB)
N.A. Callow (NAC)
D.A. Coleman (DAC)
G.A. Collins (GAC)
G.B. Collins (GBC)
L.J. Cook
G.L.A. Craw (GLAC)
J.S. Denton (JSD)
J.R. Dobson
P.J. Edwards
D. Element (DE)
S.J. Element
R. Fry
P. Goodwin
D.G. Hall
A.J. Halstead (AJH)
R.B. Hastings
R.D. Hawkins (RDH)
K. Hill
P.J. Hodge (PJH)
G. Jeffcoate
R.A. Jones (RAJ)
D. Lonsdale
P.R. Mabbott
J. McLauchlin (JMcL)
I.S. Menzies (ISM)
A.J.D. Morris
R.K.A. Morris
J. Muggleton
J.A. Owen (JAO)
M.S. Parsons
C.W. Plant
J. Pontin
C. Reid
A. Salisbury (AS)
B.M. Spooner (BMS)
J.B. Steer (JBS)
D.B. Walker
V. Wallace (VW)
M.J. Wills (MJW)
P.R. Windsor (PRW)
A.J. Wren (AJW)

# EXPLANATION OF SPECIES ACCOUNTS

Accounts are given for all the 41 species of ladybird resident in Surrey, including two colonists from the European continent that have established themselves here within the last five years or so. All the species accounts follow a similar pattern, as detailed below. Somewhat briefer accounts are included for the 13-spot Ladybird that has not been found in Surrey since the 19th Century, the 5-spot which is the only native British ladybird absent from the county, and two Australian species used for biological control that have been found out-of-doors in Surrey on very few occasions; both species have established populations in southern Europe but are not expected to survive here.

## NAME

The heading for each species account is the scientific name of the ladybird, valid throughout the world, together with the author of the name and date of its publication. The author and date are enclosed in brackets if the specific name has subsequently been transferred to another genus. Those scientific names that include numbers (such as *septempunctata*) are given in abbreviated form (*7-punctata*) with the Latin numbers replaced by numeric digits (see page 27 for the scientific names in full). English names are given for the larger species, easily recognisable as ladybirds, and are those used by the Cambridge Ladybird Survey.

## DISTRIBUTION MAP

The distribution of each species in the county is shown by a pattern of dots within the boundary of the vice-county of Surrey. Each dot represents a record from a 2-kilometre square, known as a tetrad, each of which consists of four of the small squares of the National Grid as shown on the Ordnance Survey's Landranger and Explorer series of maps. The tetrads can be visualised by considering only those grid lines with even numbers.

Each black dot denotes the presence of the species in at least one year during the period of the survey (1980-99). A few records from the year 2000 (up to June) have been added, with the justification that these specimens had overwintered and were in fact the generation of 1999. Some records from before 1980 have been included as grey dots; these all date from the 1970's and originate from a few of our recorders who were active in this decade. The dots give no indication of the abundance of the species within the tetrad but most represent permanent colonies. In the case of new arrivals (Bryony Ladybird, *Nephus quadrimaculatus* and *Rhyzobius chrysomeloides*) and species with an expanding population (Orange and Cream-streaked

Ladybirds), the distribution shown is inevitably that at the end of the survey period. The maps may overstate the abundance of migrant species which may occur irregularly almost anywhere, but perhaps only once in 20 years. The migrant species in Surrey appear to be the 11-spot and Adonis Ladybirds, although the latter has permanent colonies in the London area.

## NATIONAL STATUS

A rarity status across the whole of Great Britain has been assigned to most insects and other organisms. The statuses for ladybirds were published in Hyman (1992) and have been circulated widely with the Recorder computer database.

| | |
|---|---|
| **Endangered (RDB1)** | Occurring only as a single population or otherwise in danger of extinction. |
| **Vulnerable (RDB2)** | Declining or in vulnerable habitat and likely to become endangered in the near future. |
| **Rare (RDB3)** | Very restricted by area or by habitat or with thinly scattered populations, occurring in no more than 15 10-km squares. |
| **Notable (NA or NB)** | Uncommon in Britain with two grades of rarity, Notable A ocurring in between 16 and 30 10-km squares and Notable B ocurring in between 31 and 100 10-km squares. |
| **Local** | Limited by habitat or strongly restricted to a particular region of Britain, occurring in no more than 300 10-km squares. |
| **Recent Colonist** | New arrival since the statuses were set. |

## NUMBER OF TETRADS

This is purely a count of the black dots on the distribution map, being the number of tetrads in which the ladybird was recorded during the period of the survey.

## STATUS IN SURREY

In assigning a status to each species of ladybird in Surrey, allowance has been made for under-recording those species that are difficult to find through being extremely small or living high on trees or at ground level.

| | |
|---|---|
| **Ubiquitous** | Found almost everywhere, or at least in almost every tetrad, so distribution map shows a solid black pattern. |
| **Common** | Found in at least half the tetrads since the habitat occurs widely but not everywhere, so distribution map shows a speckled pattern. |
| **Local** | Found in less than half the tetrads because the habitat is rarer. |
| **Very local** | Found in no more than forty tetrads. |
| **Rare** | Found in no more than ten tetrads. |
| **Established** | Recently settled and with permanent colonies (localities where it can be found every year). |
| **Migrant** | Occurring regularly but without permanent colonies. |
| **No recent records** | Not recorded during the period of the survey (1980-99). |

These statuses are qualified in some way for a few of the species.

## HABITAT

This describes the place where the ladybird lives and breeds, although adults can often be found in other habitats when dispersing or in search of food. The habitat listed may be a food-plant or a host-plant for those species so restricted, or a general term such as grassland, heathland, low vegetation or trees. The latter may be coniferous or broad-leaved, and the ladybird may live among the foliage or on the trunks.

## TEXT

The text follows a broadly similar layout for each species, starting with advice on the identification of adult ladybirds, supplementing the photographs. Brief details are given on the identification of those larvae that can be recognised from their distinctive colour pattern. The life-history, habitat, behaviour and food preferences are described from observations in Surrey during the years of the survey. When other published information is used, a reference is given, which may be (*pers. comm.*) for information supplied privately.

## RECORDS

Individual records are generally given only for those species least often found, but the first few records are listed for new arrivals in the county, while unusual foods or habitats are illustrated by selected records for a few species. Each record includes the following details (if available): the name of the locality, its four-figure grid reference, a field note or brief description of the habitat, the date of the record and initials of the recorder. A reference to the publication, with volume and page numbers, is given for published records.

## *Epilachna argus* (Geoffroy, 1785) PLATE 8 **Bryony Ladybird**

(= *Henosepilachna argus*)

**National Status:** Recent colonist
**Number of Tetrads:** 16
**Status in Surrey:** Established and spreading
**Habitat:** White bryony

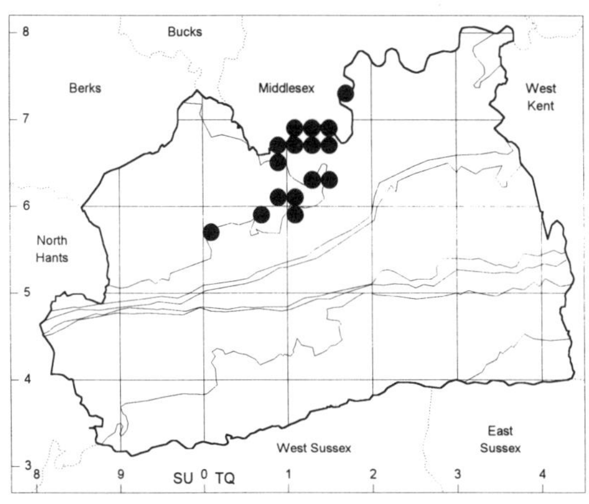


The remarkable story of the discovery of a ladybird entirely new to Britain is told by Menzies and Spooner (2000). A five-year-old child playing in a garden found a strange ladybird and kept it to show to her grandfather. Many children can tell the difference between a common ladybird and an unusual one, but Miss Alysia Menzies was fortunate to have a grandfather, Ian Menzies, with a deep interest in ladybirds and other beetles. He recognised it as something quite out of the ordinary and took it to Roger Booth, a specialist on ladybirds, who named it as *Epilachna argus*, a plant-eating ladybird from southern Europe.

This first example was found in May 1997 and was considered to be an accidental import, but a second specimen turned up in August and, when the garden and a nearby open space were searched in November, about 40 adult ladybirds were found. These were mostly beaten from ivy and honeysuckle and the ladybird was not found on white bryony, its expected food-plant, until May 1998. Meanwhile, another large breeding colony was discovered on white bryony in a nearby cemetery by Brian Spooner, quite independently and without any knowledge of the other localities.

One of the recorded sites, Molesey Heath (previously known as Field Common) was surveyed for ladybirds by Graham Collins in 1994, 1995, September 1997 and June 1998 without finding a specimen of this ladybird. He finally found one on a return visit in April 1999. These negative records suggest that the colonies around Molesey are of recent origin.

This is a large species (7 mm long) covered with very short downy hairs. The wing-cases are orange-red with eleven evenly-spaced black spots. The head, pronotum and legs are all of the same orange-red colour. The larva is pale yellow with dark spots and a tall black spine with pale side-branches on each tubercle, the spots on the central tubercles being linked. The pale yellow pupa with small black spots is covered at one end by the remains of its larval skin.

The food-plant, white bryony, is a distinctive climbing plant that clambers over hedges and fences, attaching itself by clasping tendrils that coil up to resemble a spring. It produces red berries in August before dying back to its underground stem for the winter. The ladybird feeds on the leaves, using its many-toothed mandibles to scratch away the upper surface leaving nothing but a network of veins.

The white bryony belongs to the cucumber family (Cucurbitaceae) which includes many cultivated plants: courgettes, cucumbers, marrows, melons, pumpkins and squashes. Some of these are quite frequently grown in our gardens and allotments, and the ladybird is capable of feeding on their leaves; indeed it has been reported as attacking melons in other parts of Europe. However, all our observations to date suggest that the newly-established population around Molesey is feeding solely on white bryony.

Although this is a warmth-loving species from the Mediterranean, it has been spreading northwards in Europe over many years. A small colony in eastern Germany, quite small in 1954, had expanded to cover an area of 80 by 110 km in 1976 (Klausnitzer, 1997). This kind of expansion may be expected here, particularly when the ladybird reaches the areas of chalky and sandy soil on which white bryony is most common.

Further colonies were found in 1999 in Cobham, 8 km to the south-west of the Molesey garden, initially by David Baldock and later by myself. Although the ladybird was feeding on three well-separated plants of white bryony, many others were not attacked, nor were other plants in neighbouring towns and in many other parts of Surrey.

A further spread outwards was reported by Brian Spooner in May and June 2000, including St. Mary's Church, Walton, where the ladybird had been absent in 1998, and sites in East Molesey and around Esher, although I had searched both these areas in 1999 without finding it. Several new records were on isolated plants of white bryony which the species seems well able to locate.

Also in June 2000, another new colony was found by David Baldock at Ham, 5 km north-east of the East Molesey locality and across a loop of the River Thames. White bryony was growing over the perimeter fence of some allotments in two places and was recently into flower. It was teeming with ladybirds including mating pairs.

This species has also appeared under the name of *Henosepilachna argus* (Fürsch, 1967; Menzies and Spooner, 2000) but dissection of specimens from West Molesey has shown that it is not a typical *Henosepilachna* but intermediate to *Epilachna* in its characters. The present split between *Henosepilachna* and *Epilachna* is therefore unsatisfactory and the species is best called *Epilachna argus* for the moment (Roger Booth, *pers. comm.*).

RECORDS: **West Molesey, High Street** (TQ1367), in garden, 14.5.97, on climbing frame, 1.8.97, nine beaten from ivy and honeysuckle over garden fence, 1.11.97, adult in flight, 29.4.98, 20 adults on white bryony, 31.5.98 (Alysia Menzies & ISM); **Molesey Heath** (TQ1367), beaten from ivy, 7.11-13.12.97 (ISM with others), many subsequent records; **West Molesey, Elmbridge Cemetery** (TQ1368), large numbers feeding on white bryony, with adults, eggs, larvae and pupae all present, 22.5.98 (BMS), many subsequent records; **Walton to Molesey** (TQ1067-1368), on many plants of white bryony at various localities along the Thames, by A3050 road and near reservoirs, 11.7.98 (BMS); **Cobham, near Sainsbury's** (TQ0960), on white bryony, 6.99 (DWB); **Cobham, Between Streets** (TQ1060), beaten from white bryony, 1.9.99 (RDH); **Cobham, Church Path** (TQ1059), beaten from ivy with white bryony, 1.9.99 (RDH); **Walton, public park** (TQ0965), 31.5.00 (BMS); **Walton Bridge** (TQ0966), 31.5.00 (BMS); **Walton, St. Mary's Church** (TQ1066), 31.5.00 (BMS); **East Molesey** (TQ1468), 3.6.00 (BMS); **East Molesey, south of River Mole** (TQ1467), 3.6.00

(BMS); **Arbrook Common** (TQ1463), 4.6.00 (BMS); **West End Common, Esher** (TQ1263), 4.6.00 (BMS); **Ham, opposite Ham House** (TQ1772), many on white bryony, including mating pairs, 7.6.00 (DWB); **Woking Park** (TQ0057), 8.7.00 (AJH); **Wisley, RHS Garden** (TQ0658), very light infestation, 10.7.00 (AJH).

## *Subcoccinella 24-punctata* (Linnaeus, 1758) PLATE 8 **24-spot Ladybird**

**National Status:** Common
**Number of Tetrads:** 379
**Status in Surrey:** Ubiquitous
**Habitat:** Grassland, especially roadside verges

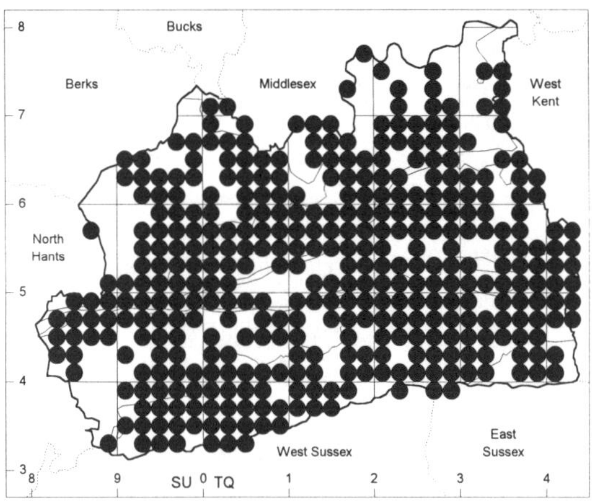


Before the arrival in this country of the Bryony Ladybird, the 24-spot was our only vegetarian species feeding directly on the leaves of plants. It is not often noticed but can be found in profusion among long grass on roadside verges and in similar places, although it is generally absent from heathland and uncommon in built-up areas. It is a rather small (3.5 to 4 mm) orange-red ladybird with many round or oval black spots that tend to join up into irregular blotches, although specimens with 24 free spots can also be found. The insect is covered with very short hairs which can just be seen with a 10x lens when they catch the light. The underside and legs are the same orange-red colour as the wing-cases but the pronotum is slightly contrasting in colour, being brownish-red and often with a dark central spot.

Few people know the larva of the 24-spot Ladybird but it is quite distinctive and easy to find. It is pale creamy-yellow in colour, short and stubby in shape and covered with a thicket of spiny bristles, since each tubercle on the abdomen is topped with a tall spine which itself gives off many strong side-bristles.

At the start of this survey, the British population of this ladybird was known to feed mainly on the leaves of various species of campion, principally red campion in south-east England, although the species could be a pest on lucerne, a fodder crop, in south-eastern Europe (Marriner, 1927; Richards, Pope & Eastop, 1976). In 1985 and 1986 I found larvae feeding on the leaves of the red, white and bladder campions, all native species, and also on the related soapwort, an escape from gardens. Most of these were single observations and searching campions produced very few records.

Meanwhile Arthur Baldwin of Kew, which lies just inside our survey area, had been studying aspects of the biology of the 24-spot Ladybird using captive breeding stock which he fed mostly on chickweed. He noticed that wild specimens were almost all breeding on the false oat-grass, then moving to nettles and thistles in the autumn. The ladybirds did not eat the leaves of nettles, but fed on the flowering parts, and were not at first noticed to feed on the thistle either (Baldwin, 1988, 1990). False oat is a tall grass which is generally abundant and is dominant on many roadside verges, particularly those that are cut annually with the cuttings left to rot down.

These observations have since been repeated and extended to cover the whole of the county. The presence of false oat has proved to be a very reliable indicator for the presence of the ladybird. Feeding on the leaves of this grass has been both observed directly and inferred from the characteristic grazing marks on the grass leaves. The upper surface of the leaf and the green material below it is eaten away, leaving behind the colourless lower surface, so the feeding mark shows through as a pale patch between the veins. These grazing marks may run along the leaf, or transversely or diagonally across it. This feeding on the upper surface of grass leaves is in contrast to the feeding on campion reported by the authors cited above, which is from the lower surface.

| | larvae | adults |
|---|---|---|
| observed feeding on false oat | 3 | 5 |
| grazing marks, with ladybirds present | 6 | 6 |
| on false oat, not seen feeding | 8 | 55 |
| on false oat mixed with other plants | 1 | 92 |

***Number of records of 24-spot Ladybird on false oat***

It is remarkable that other tall, coarse grasses occupying the same habitat as the false oat are not used. In grassy areas without this species, such as the Surrey Wildlife Trust's reserve of Bagmoor Common where a large area of acid grassland is dominated by purple moor-grass and wavy hair-grass, this ladybird is totally absent. I have also examined cock's-foot, tall fescue and various species of meadow-grass and bent, and found no evidence of feeding, although a few ladybirds were present in tufts of cock's-foot grass in late October, presumably overwintering.

A careful search was made for feeding on other species of grass and eventually one was discovered, the soft brome. The few records of this are listed below. Arthur Baldwin also reported feeding on nettle and thistle, although false oat was preferred. Most of our records from September and October come from grass tufts, but many others from thistles, especially creeping thistle, and a few from nettles. Ladybirds probably need little food at this time of year and I have found no other evidence of the species feeding on these plants except for a larva on spear thistle, with no grass nearby, at Farthing Downs on 4 August 1999. It is quite likely that some other food-plants of minor importance have been overlooked, particularly those on which the ladybirds feed from beneath the leaves.

Our distribution map has become essentially one of the 24-spot associated with false oat, and its previously recorded association with campions was largely forgotten. At the end of the survey I searched for evidence that it still fed on campions and eventually managed to find some ladybirds on red campion and one plant of white campion with grazing marks on its leaves. The other traditional food-plant, lucerne, is rarely grown as a crop in Britain but comes up occasionally on roadside verges sown with imported seed. A brief glance at some plants beside the A3 at Cobham in 1999 showed that just one out of several patches of lucerne had two ladybirds on it, and possibly also grazing marks on the leaves.

Studies of ladybirds around Kew in west London from 1963 to 1974 have already shown that this species only has one generation a year in Surrey (Richards, Pope & Eastop, 1976). During the present survey, mating pairs were noted in April and May, and larvae from the second half of May to the beginning of September (see charts). Two further mating pairs, seen at Bagshot and Chobham on 28 September 1997, would be too late in the year to give rise to another generation and can be put down to confusion of the seasons, perhaps caused by a cold spell followed by a mild one.

The most recent summary of the biology of ladybirds in Europe (Klausnitzer, 1997) lists food-plants from various families but no grasses at all, although Tanasijevic (1958) lists several species of grass from a study in Yugoslavia, but not the false oat, and Majerus (1994) lists grasses in general among many other plants. A closely-related species, *Cynegetis impunctata*, feeds on grass on the Continent and it has been necessary to examine a few specimens carefully to confirm that our grass-feeding 24-spot Ladybird really is that species. The ladybird certainly has benefited by adapting to the false oat, for this is an abundant species, while lucerne is a rare casual and campions, although widespread and common, are comparatively hard to find.

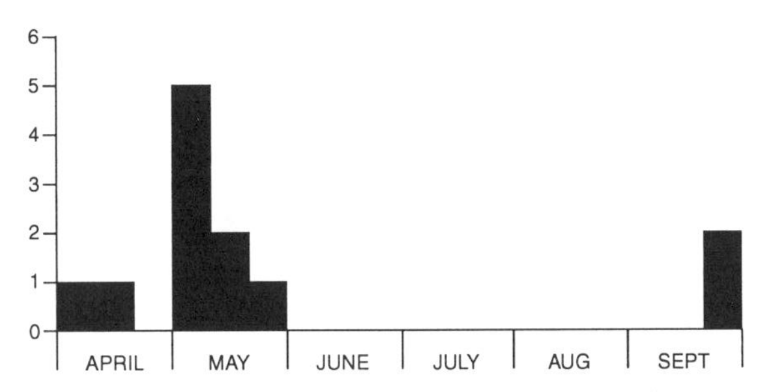


***24-spot Ladybird – number of mating pairs***

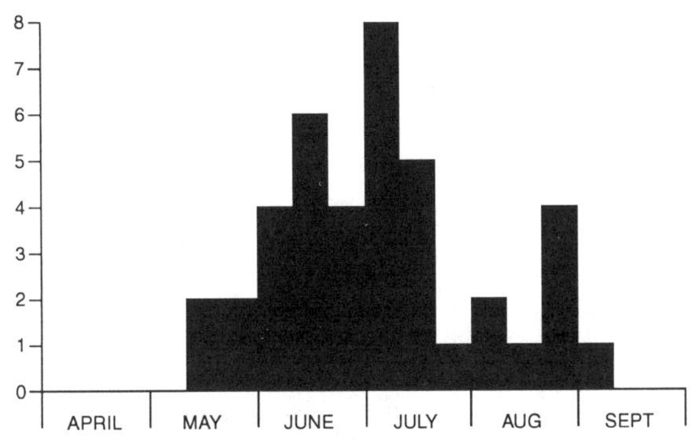


***24-spot Ladybird – number of larvae***

The conclusion that the 24-spot has built up a large population through feeding on a species of grass that was not recognised as a food-plant before the start of this survey, either in Britain or anywhere in Europe, is remarkable. Perhaps the food-plants of this species should be reassessed, both elsewhere in Britain and in nearby parts of the Continent. The larvae are easy enough to find, but care should be taken in interpreting the grazing marks, since yellow spots on grass leaves may have other causes.

SELECTED RECORDS: **Horley** (TQ2941), four adults, including mating pair, on leaves of soft brome with feeding marks, 6.5.95 (RDH); **Hutchinson's Bank, Addington** (TQ3861), adults, larvae and feeding marks on soft brome, 17.5.97 (RDH); **Riddlesdown Quarry** (TQ3359), adult on false oat-grass with feeding marks; soft brome also has feeding marks on leaves, 16.6.97 (RDH); **Banstead Downs** (TQ2561), one specimen with absolutely no spots on wing-cases or pronotum, among many typical examples, 13.5.00 (GAC).

## *Chilocorus bipustulatus* (Linnaeus, 1758) PLATE 11 **Heather Ladybird**

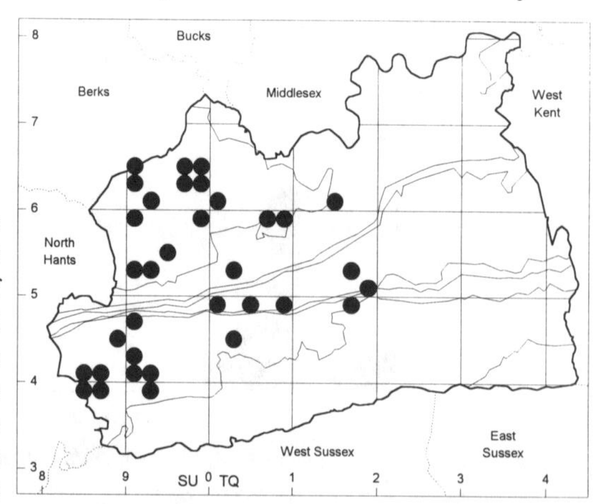


**National Status:** Local
**Number of Tetrads:** 34
**Status in Surrey:** Local
**Habitat:** Heathland; also on juniper

The Heather Ladybird resembles a small version of the Kidney-spot Ladybird. It is shiny black, of similar rounded shape with head drawn in and with wing-cases splayed at the outer edges, but the central red spot is replaced by a transverse line, usually broken into several short dashes. It is a widespread European species feeding principally on scale-insects in various habitats, but in Britain it is associated chiefly with the common heather, hence its name. The mobile adults are also often found on the young pines, birch, oak and other plants that share its heathland habitat. We have, however, no records from pines away from heathland.

The larva is dark grey with a pale central stripe. On each tubercle there is a tall bristly spine which is generally black. A whitish band across the whole of the first abdominal segment, including the spines, distinguishes it from its relatives, the Pine and Kidney-spot Ladybirds.

This is rather a common species on the heathlands of west Surrey but records from elsewhere are very sparse. Since 1996 we have recorded it from some of the few remaining sites for native juniper along the North Downs: Box Hill, Hackhurst Downs and Newlands Corner. The possibility that it may also occur in gardens is suggested by two records from abandoned nurseries where the ladybird was found on cypresses, which are closely related to juniper. It certainly breeds on the juniper, since larvae were present in July at Box Hill. This species was not found in a comprehensive study of the fauna of juniper on the North Downs by Dr Lena Ward in 1968, but it may be significant that an introduced species of scale-insect, *Carulaspis juniperi*, was found on native juniper for the first time in that survey, and at all our three sites for the ladybird (Ward, 1970, 1977). It is unfortunate that none of our records involves feeding.

Just occasionally the Heather Ladybird turns up on the trunks of ash trees, a habitat favoured by its two relatives, the Pine and Kidney-spot Ladybirds. At Box Hill in 1989 one was seen walking slowly up and across an ash trunk, moving its palps; it walked past a young oyster scale (*Chionaspis salicis*) and ignored it, although a Kidney-spot Ladybird was feeding on these scale-insects on a nearby trunk.

We have too few records (90 specimens) to make much comment on the life-history of this species; their dates tend to be those when recorders found it convenient to visit the heathlands of west Surrey. Adults were seen in spring until early May, larvae were only noted on 20 July and the new generation had most records from the second half of August, but adults found on 9 & 28 July are difficult to assign to a generation. Those records are listed that give more information than simply being beaten from pine, or swept from or found on heather.

SELECTED RECORDS: **Blackheath, near Guildford** (TQ0345), 6 on Scots pine, walking slowly along shoots and needles, 19.8.84 (RDH); **Burpham** (TQ0252), 3 on or inside open dry cones on a group of small upright cultivars of an ornamental cypress (*Chamaecyparis* sp.) in abandoned nursery (scale-insects present, *Juniperus communis* nearby), 9.3.85 (RDH); **Box Hill** (TQ1752), on ash trunk (observations in text), 26.8.89; (TQ1851), beat 4 larvae from single juniper bush, 20.7.96, beat adult from juniper, 31.3.97 (RDH); **Dorking** (TQ1749), beaten from small Lawson's cypress in abandoned nursery, 1.5.95 (RDH); **Mare Hill Common** (SU9239), on woody stem of heather, 5.5.96 (RDH); **Hackhurst Downs** (TQ0948), on juniper, 6.9.96 (GAC), beaten from juniper, 30.3.97 (RDH), on juniper, 12.10.98 (GAC); **Warren Farm, Guildford** (TQ0149), on beech trunk, apparently searching for food, 1.4.97 (RDH); **Newlands Corner** (TQ0149), 2 beaten from juniper, 1 visible on bark of twig, 1 active on trunk, 1 resting on needles, 1.4.97 (RDH).

## *Chilocorus renipustulatus* (Scriba, 1791) PLATES 12, 14 **Kidney-spot Ladybird**

**National Status:** Common
**Number of Tetrads:** 228
**Status in Surrey:** Common
**Habitat:** Trunks of sallow, ash, alder and occasionally birch

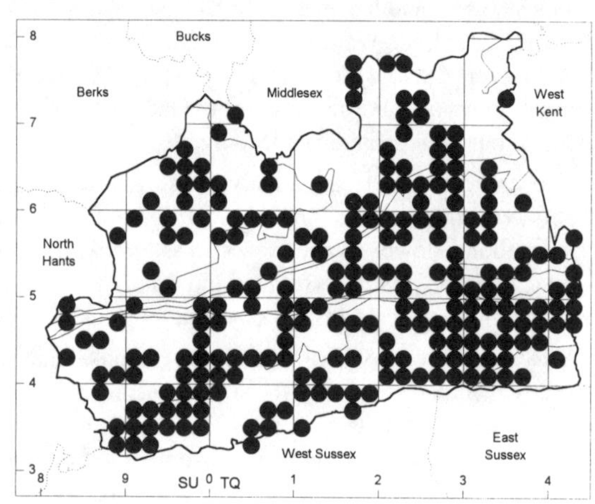


The Kidney-spot is one of those large ladybirds of distinctive pattern which attract the attention of all who see them, however disinterested in insects they may otherwise be. With its shiny appearance, hump-backed shape with tucked-in head, and wing-cases turned outwards at the edges, it is totally unmistakable. The two red spots on a black background are reminiscent of a 2-spot Ladybird with the colours reversed. Because of this, it is often assumed by a beginner to be a form of the 2-spot, but the melanic variety of the latter species has generally either four or six red spots and the central area of the wing-cases is always black. The large red spot on each wing-case is oval and sometimes indented at one side, a shape that has given rise to both the scientific name *renipustulatus* and the English name. Vegetarians may prefer to regard this shape as that of a broad bean. The adult ladybird is ponderous and slow-moving, especially when compared with the frenetic activity of the Pine Ladybird which often shares its habitat. The Kidney-spot is a slightly larger species at 4 to 5 mm long, and appears more hump-backed when seen on a tree trunk.

The small, dark grey larva is more squarely oval and stubby in appearance than the larva of the Pine Ladybird with which it is often found. Its body colour is reddish brown after a moult but then turns blackish. Each tubercle has a tall black spine, hardly branched but with side bristles. The larva is distinguished from those of the Heather and Pine Ladybirds by the absence of any white colour on the first abdominal segment. The black spines may appear to have some superficial white specks, but these are bits from the scales on which it has been feeding.

The ladybird is a specialist feeder on scale-insects, also known as coccids, and is found principally on the trunks of ash and sallow, less frequently on alder, birch and some other broad-leaved trees and shrubs. Here the term 'sallow' is used as a general name for small, often shrubby, broad-leaved willow trees, principally the goat willow, *Salix caprea*, and the grey willow, *Salix cinerea*. It is easier for an entomologist to use the general term than to name the trees to species, especially since the diagnostic leaves may not be out when the ladybirds are seen on the trunks. We have in fact found Kidney-spot Ladybirds, both as adults and larvae, on the trunks and branches of both goat willow and grey willow.

A very common scale-insect occurs widely on all the trees where the ladybird is found, but perhaps chiefly on sallow. This is *Chionaspis salicis*, popularly known as willow scale or oyster scale, which is described by Newstead (1901). It has been known for 50 years as the principal food of the Kidney-spot Ladybird following studies in Germany (Klausnitzer, 1997). An even more common scale-insect on ash trees is the felted ash scale *Pseudochermes fraxini* (see page 52). The scale of this insect resembles a tiny piece of cotton-wool, but the oyster scale is recognised by its firm scale with the minute brown larval skin remaining at one end. The males have an elongate white scale while that of the female is greyish-white and eventually becomes curved into the shape and appearance of an oyster. The ladybirds are regularly found among colonies of one or the other of these scale-insects on ash trunks, and of *C. salicis* on sallow and occasionally other trees. I have also seen both the ladybird adult and larva feeding on each type of scale-insect.

On 30 June 1984 a larva was eating a little pale orangey mobile coccid on an ash trunk among a colony of *C. salicis*. Because of the date, I take it to be the male of this species. Newstead wrote that there was often a great preponderance of males, whose scales he illustrated almost covering a branch of ash, and that the active insect emerged from the pupa beneath its scale over about two weeks from the end of June onwards. Masses of empty scales of male *C. salicis* were found on ash in South Croydon on 30 July 1984, accompanied by three adults, some larvae and four separate clusters of about 15 shed larval skins of the ladybird, but the only two pupae found were under leaves. An ash tree near Dorking was covered with active reddish male *C. salicis* and their empty scales on 15 July 1990. Among them were over 20 larvae of the Kidney-spot Ladybird and clusters of cast-off skins. The clustering of larvae, pupae and shed larval skins has been reported previously, although the link with mass occurrences of male *C. salicis* has apparently not hitherto been recognised. Extensive colonies of larvae (one of 170 individuals) were found upon the larger limbs of five ancient ash trees in Colchester, Essex, on 23 June 1945 (Cox, 1946). An aggregation of about 45 pupal cases on an ash trunk is illustrated by Majerus (1994, page 67). Such clustering of pupae has been seen only once during this survey, with a cluster of 12 and a few other pupae, larvae and adults on a small ash at Farnham on 4 September 1988.

In spring most adults are found in April and May, with some surviving into June, but from then on it is difficult to distinguish the few overwintered individuals from early emergences of the next generation. We have only twice found mating pairs in spring, on 15 April 1989 and 13 April 1996. A mating pair in autumn, on 23 September 1997, is likely to have mistaken the season. One appeared to be laying eggs on 21 May 1989 but I have seen this more precisely in another county, when the ladybird inserted its egg beneath the scale of a coccid. Larvae can be found from June onwards and are most numerous in mid-July, but a few specimens are still larvae in early September. Although larvae, when feeding, are invariably found on the trunks and branches, the great majority of pupae found in this survey (42 out of 65) were on or under leaves. Most pupae were recorded between mid-July and early August. In one colony examined in 1986, some adults were emerging but most were still pupae on 22 July, while most if not all had emerged by 13 August. Two individuals were dated precisely. A fresh pupa on a birch leaf, still yellowish in colour when found on 20 July 1986 (later turning black), emerged on 31 July. A larva, found on a sallow leaf on 23 July 1988, pupated on 25 July and emerged as an adult on 6 August. If keeping these pupae in captivity did not affect their development, this puts the pupal stage at 11 to 12 days.

The distribution in Surrey of the Kidney-spot Ladybird is curiously patchy. There are few records from the built-up areas of London or from the heaths of west Surrey, where those found were on birch and sallow. It must of course follow the scale-insects on which it feeds. It is widespread on sallow but in some areas the sallows have neither scale-insects nor ladybirds on them. It is often found on ash, but some lichen-covered ashes have no scales and no ladybirds, while others support only the Pine Ladybird. Although the Kidney-spot has been seen eating the felted ash scale, it is possible that it only occurs on ash when oyster scale is present.

Some indication of the relative abundance of the Kidney-spot Ladybird on its different hosts is given by the number of specimens reported to this survey – 700 on ash, 320 on sallow, 19 on alder, 14 on lime, 9 on birch and 3 on field maple; *Chionaspis salicis* was also noted on each of these trees. The ladybird was twice found on lime trees infested with the introduced *Pulvinaria regalis* (see page 54) as well as by its usual prey, *C. salicis*, but with no evidence of feeding on the former insect. It is noteworthy that our largest coccinellid predator of scale-insects has failed to tackle this new pest, not even feeding on its eggs as does the Pine Ladybird. Other occurrences of Kidney-spot Ladybird are considered to be casual, except for five found on various dates on the trunks of ornamental species of *Sorbus*, which may harbour suitable prey, and four beaten from native juniper at Newlands Corner on 18 September 1994. Another introduced scale-insect, *Carulaspis juniperi*, has spread from cultivated junipers to the native plant (Ward, 1970). The ladybirds may have been feeding on this, but the following record is more definite.

A colony was found in Horley in 1984 on a coniferous bush in a garden. A scale-leaved cultivar of *Juniperus*, not precisely identified, was under attack from a coccid with a firm white scale, later named as *Carulaspis juniperi* by Dr Jennifer Cox, then of the Natural History Museum. Up to eight Kidney-spot Ladybirds were on this bush from 12 April to 4 June and were seen to feed on the scale-insect. The colony was permanent, since among three adults on 18 July was one newly-emerged near an empty pupal skin. Ladybirds were then present until at least 26 September, with a maximum of six on 2 August, but in October the gardener removed and destroyed the juniper bush with its many dead brown sprays, bringing an unfortunate end to what I considered as an experiment in biological control.

Our understanding of the habits of this species was held up for years by an unfortunate mistranslation from the French by Fowler (1889). He wrote that the larvae of *Chilocorus* feed, according to Mulsant, on gall insects, but he mistook the word *gale*, a scab or scale, for *galle*, a gall. Fowler gave no association apart from woods and hedges, but Joy (1932) and Pope (1953) list it as chiefly occurring on sallow. The link with ash goes back at least 50 years to the clusters of larvae found in Essex in 1945, as mentioned above, and a report from Bedfordshire of many resting motionless on young stems of ash, high above the ground, on 14 April 1946 (Verdcourt, 1952). Crowson (1956) appears to be the only author of this period who wrote that its occurrence on ash was normal and to be expected. My own first encounter with this species was in Hertfordshire on 24 August 1975 when several were present on the uppermost trunk of a tall ash cut down by a conservation working party. Because Joy said that it occurred chiefly on sallow, we assumed that it had migrated from a bush of this type occurring nearby, a case of believing the textbook rather than the evidence of one's own eyes.

## *Exochomus quadripustulatus* (Linnaeus, 1758) PLATES 12, 14 **Pine Ladybird**

**National Status:** Common
**Number of Tetrads:** 448
**Status in Surrey:** Ubiquitous
**Habitat:** Trunks and branches of pine and other conifers, ash, sycamore and other broad-leaved trees

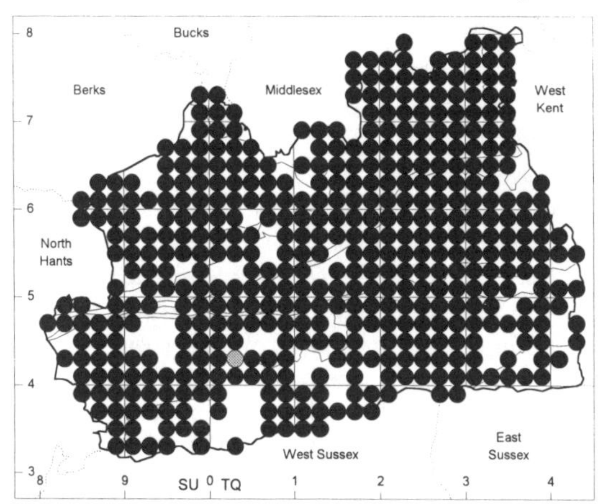


The Pine Ladybird is one of our most abundant species but goes largely unnoticed because of its dark colour and special habitat. It lives on the trunks, branches and small twigs of trees, rather than among the foliage. Most people will have a colony of this ladybird near their home without realising it. It is the commonest ladybird on pine trees and one hundred years ago was known almost exclusively from pines and perhaps some other conifers. This was stated in most books about beetles and gave rise to the name of Pine Ladybird. About fifty years ago it was first noticed on the trunks of ash trees and this population now rivals and even exceeds that on conifers, at least in Surrey. More recently a further large population has built up on the trunks of other broad-leaved trees, principally sycamore and other maples, horse-chestnut and lime, where the ladybird is feeding on the egg-masses of an introduced scale-insect.

There are nearly always four red spots on the black wing-cases but these can be small and sometimes dull, so the initial impression is of a black ladybird, especially when seen from a distance. The foremost spots always have a tail extending around the shoulder of the wing-case, thus forming a characteristic comma-shape. The size of the ladybird varies markedly from 3 to 4.5 mm – usually the males are smaller. They cannot be confused with the two smaller ladybirds having four red spots since these have a covering of downy hairs, while the Pine Ladybird is shiny and almost hairless. The only possible confusion is with melanic examples of the 2-spot, in which the red colour can be reduced to four spots of very much the same shape and in the same position as the spots of the Pine Ladybird. The most obvious distinction is that the wing-cases of the 2-spot go straight down at the sides whilst those of the Pine Ladybird splay out to form a narrow flange which is characteristic of the small group of species living on tree trunks. The pronotum of the Pine Ladybird slopes steeply forwards and the head is held below it, creating a more rounded shape overall than is usual among British ladybirds. The underside and legs are black with some red colour at the tip of the abdomen which is lacking in the 2-spot.

The larva appears at first sight to be dark grey with a white spot on each side. On closer examination, the body-colour is fawn or light grey with a pale central stripe, while all the tubercles have a tall black spine with short side-branches ending in a pale bristle, except for the middle tubercle on each side of the first abdominal segment and usually the one below it, which are creamy-white. The larval skin splits open on pupating but remains in place, so that the pupa can be identified from the characters of the surrounding larval skin. The pupa is red at first but soon turns black.

Around 1980 I became intrigued by the sight of black ladybirds running about on the trunks of ash trees, but it was not until 1983 when I met a ladybird specialist, John Muggleton, that it was confirmed that these were indeed Pine Ladybirds, but in an unexpected habitat. Further observation eventually showed that there were permanent colonies on the ash trunks, and both adults and larvae were feeding on a tiny red insect living in cracks in the bark, attached to the tree and covered with white waxy 'wool'. This is a scale-insect named *Pseudochermes fraxini* (Kaltenbach), or the felted ash scale, and is described by Newstead (1903). He had discovered it as a new species in 1891 and named it *fraxini*, but was anticipated by Kaltenbach who had found it in Germany and, by happy coincidence, also named it *fraxini*. Newstead (1901) wrote that it was apparently immune from the attacks of both birds and insects. If only he could have lived another hundred years! During this survey, over 2500 specimens of Pine Ladybird have been seen on ash trees, mostly among felted scales, and observed feeding on these scales on 16 occasions, both as adults and as larvae. The Victorian entomologists were observant and travelled widely, so it is unlikely that such a conspicuous habit could have been overlooked. The earliest reference I can find to Pine Ladybirds living on ash is from a wood in Bedfordshire on 22 September 1946, when some were found on and beneath ash trees (Verdcourt, 1952). From 1984 to 2000 they have been abundant on ash almost everywhere in Surrey that the tree occurs, and are most often seen on young trees where the smooth bark is just starting to break up into cracks.

The life-history of the Pine Ladybird can be illustrated using principally the population on ash trees, which was observed in great detail, particularly in 1984. Although the ladybirds are active and begin mating in early spring, development is slow and only a single generation is produced, with a clear gap separating the new generation from that of the previous year. This species becomes active very early in the year, in fact appearing on the tree trunks at the first hint of warmth or winter sunshine, but we have no records of activity later in the year than 6 December 1986 or earlier than 30 January 1995. They appear so quickly that their overwintering quarters must be on the trees. Even though active and feeding low on the trunks in the daytime, ladybirds can often be seen moving upwards as the sun goes down. The majority of suitable sites for hibernation are out of reach, but inactive ladybirds found between November and February were in bark crevices or under flaps of bark on trunks of ash and oak, by the terminal bud of an ash shoot, or clustered around the end buds or the base of side shoots on small pines. Mating pairs are frequently seen on the trunks in April, or in March if the weather is favourable, and the mating season extends into May and sometimes just into early June. Eggs appear to be laid into cracks in the bark, but this has not been observed closely. In 1984 the last pair was seen on 30 May and the last surviving adult on 15 June. The first larvae were also noticed on this latter date and then observed regularly until the last of them pupated on 4 August. Pupae were noted from 11 July to 31 August, mostly on the trunks but with one under a leaf, and the first fresh adults were out on 22 July. Two individuals, which had just pupated on 19 July and were still there on 30 July, had emerged and gone when checked on 1 August and 2 August respectively, giving the duration of the pupal stage as 12 to 13 days. When just emerged, the head and pronotum are black but the wing-cases are yellow, then red and finally dark brown with orangey spots. The ladybirds remain this colour for a considerable time. The fresh adults are generally inactive in August and absent from the ash trunks, probably resting higher on the tree; one was found by a terminal bud. In 1984 they returned to the trunks in numbers on 16 September and other sunny autumn days. A mating pair was seen on 16 September 1984 and another on pine on 9 September 1995, but I ascribe this merely to a premature feeling of spring. The pattern of 1984 was repeated in subsequent years with some variation in dates, but always with slow development giving a clear gap between the generations in early July.

The Pine Ladybird lives up to its name in Surrey and is abundant on Scots pine wherever this tree is found, whether invading heathland, growing in plantations or as isolated amenity trees. The first ladybird on newly planted pines is usually a 7-spot, but the Pine Ladybird is generally the first of the conifer specialists to arrive. We have records from pines in 118 tetrads, showing it to be more widespread than any of the other pine species. In early spring the ladybirds are numerous and active on the trunks of pines, but later on they frequent the branches and small twigs. Both adults and larvae are often found among colonies of woolly aphids, not only on Scots pine but also on other conifers, especially Douglas fir, Norway spruce and larch. Our records list 820 specimens on Scots pine, 100 on Douglas fir, 80 on Norway spruce, 40 on larch, 12 on black pine, a mating pair on Weymouth pine, a few adults on lodgepole pine and various exotic spruces, and larvae among woolly aphids on silver fir on one occasion. Both cedar and western hemlock appeared to be free of the pests on which this ladybird feeds, but eventually a few adult ladybirds were found on these trees as well.

Some minor hosts may be overlooked by comparison with the large numbers referred to above. Every year some Pine Ladybirds are found on oak, mostly in spring and autumn and totalling over 100 specimens. Several scale-insects are known from oak but I have found neither larvae nor pupae of the ladybird on this tree (except for one pupa in close proximity to an ash). Some of the blank spaces on the distribution map are country areas on clay soils where the trees are predominantly oaks and this ladybird is very difficult to find.

A second species of scale-insect lives on the trunks of ash trees, the oyster scale *Chionaspis salicis* that is perhaps more frequent on sallow (see page 49). Both *Pseudochermes fraxini* on ash and *C. salicis* on sallow are given as essential food items for the Pine Ladybird by Mills (1981), meaning that he found larvae of the ladybird feeding on these insects in the wild. Most observations of Pine Ladybird on ash trees in this survey have been among felted scales, but some have been seen among oyster scales on at least two occasions. A few specimens, 12 in total, have been found on sallow over the years, mostly adults on trunks in the spring and on trunks and under leaves in the autumn, but two pupal cases were found on *Salix caprea* at Chobham on 23 August 1987. This confirms that sallow is a host for the ladybird in Surrey, though only a minor one, with *C. salicis* the likely prey.

A rather similar insect to the felted ash scale occurs on trunks of beech. This is *Cryptococcus fagi*, the felted beech scale, which is described by Newstead (1903). The tiny yellow females live in crevices in the bark, and successive generations produce masses of white felted material on old trees. This is another species that Newstead (1901) considered to be immune from predation by birds and insects. We have records of over 50 Pine Ladybirds, including mating pairs and larvae, found among these scale-insects on which they are known to feed (David Lonsdale, *pers. comm.*).

Adult Pine Ladybirds have been beaten from bushes of native juniper at five of its few remaining sites in Surrey, and also a larva at Brockham Quarry on 10 June 1990. The probable prey is an introduced scale-insect, *Carulaspis juniperi*, that has spread onto native juniper from plants grown in gardens (Ward, 1970). This insect has a more solid scale-like covering than those mentioned above. Occasional ladybirds found on Lawson's cypress and other garden conifers may be feeding on this scale-insect or a similar one.

In almost every year one or two Pine Ladybirds are found on gorse, some overwintering among the spines but others later in the year, including a mating pair. There are scale-insects on gorse but our only observation is of the ladybird apparently eating a black aphid. A few ladybirds have also been found on the woody stems of broom, including one bush with scale-insects at the base of its twigs.

Two adult Pine Ladybirds were beaten from an apple tree at Molesey on 19 August 1998. There was copious white fluff on the foliage, probably produced by the aphid *Eriosoma lanigerum*, and one ladybird was covered with white threads. At Earlswood on 14 April 1985, seven Pine Ladybirds including two mating pairs were on the trunk of a small ornamental tree, probably a cherry, in the presence of tiny white shield-shaped scale-insects; this observation has not been followed up.

Beating three Pine Ladybirds from yew at Bookham Common on 29 June 1996 seemed at first of little importance, but I could not ignore the succession of pupae appearing on walls outside my house in Horley during the weeks that followed. These were traced to a small bushy yew with a brown globular scale-insect on its twigs. This is likely to be *Parthenolecanium pomeranicum*, formerly known as *P. taxi* (Andrew Halstead, *pers. comm.*). Six ladybird pupae were found on the walls of the house and many more pupae and larvae on the yew bush. In this type of scale-insect, the eggs are protected by the body of the dead female, but without a white waxy covering.

About 40 years ago an extremely large scale-insect was found for the first time in Britain on street trees in the London area. This is *Pulvinaria regalis*, a native of South America which at 7 mm long is a giant among the tiny insects of this family. It has been given the name of horse-chestnut scale but it also attacks other trees, principally lime and various maples. It is mobile and lives initially under the leaves of its host, then overwinters on the young twigs before descending to the main trunk to lay its eggs and die. The eggs are covered with a mass of white waxy wool formed beneath the body of the female. When in profusion this makes the tree unsightly and has aroused considerable comment. My first record of an association with Pine Ladybird was an unhatched pupa among the previous year's egg-masses of *Pulvinaria* on a Norway maple in Croydon in April 1985, followed by 13 empty pupal cases on a lime tree, recently colonised by the scale-insect, in Horley in November 1985. During the years that followed, it became clear that the ladybird had adapted to this new food in large numbers. I was able to observe this through much of the year on horse-chestnuts in Nonsuch Park in 1993. In March many ladybirds were active and some were mating on the trunks near the old egg-masses, while there were none among the live half-grown scale-insects along the twigs. By May the scale-insects had descended and the fresh egg-masses were covering the trunks, with many ladybirds near them, some still mating. On 13 June there were over 100 ladybird larvae with heads inserted into the egg-masses, or near them with heads covered in white fluff. The dead body of the old female scale-insect was often detached and left dangling. Larvae were seen with their mouthparts working on the white material, but it was not quite clear whether they were removing it or actually eating it. Many tiny yellowish larvae of the scale-insect were moving about slowly, sometimes actually walking on the ladybird larvae, but were completely ignored. The ladybirds continued to feed on the egg-masses and pupated near them in large numbers. Our records of Pine Ladybird associated with *Pulvinaria* list 650 on sycamore, 180 on horse-chestnut, 160 on Norway maple, 100 on lime, and smaller numbers on elm, silver maple and box elder (an American maple). Pine Ladybirds are rarely found on any of these trees when not attacked by *Pulvinaria.*

The attentive and persistent reader will have noticed from the above that Pine Ladybirds almost always find their prey beneath a protective covering of waxy white wool, so the question arises – is this covering matter itself the food? Superficially it seems little different from the powdery mildew eaten by other ladybirds. The idea is supported by the larvae continuing to feed on the egg-masses of *Pulvinaria* after the eggs have hatched, and by a further strange observation. On ash trees in spring where this species has been feeding, some scale-insects can be found with the waxy covering removed but without any other damage (I have seen this once resulting directly from nibbling by a Pine Ladybird). When Kidney-spot Ladybirds are present, the bodies of the scale-insects have been attacked and show as a red blotch against the white covering.

An experiment designed to settle this point proved somewhat inconclusive. Two hungry Pine Ladybirds were introduced to two recently dead *Pulvinaria* that had been persuaded to lay their egg-masses on a small twig. The resulting behaviour was observed under a binocular microscope for four hours over two evenings. The ladybirds nibbled at the edges of a dead scale-insect and strained to lift it. Once removed, it was ignored. They plunged into the mass of white waxy wool but it could not be determined whether they were feeding on this or the eggs, since the mouthparts of the ladybird lie beneath the head. It almost certainly chews the eggs, since bits of eggshell were found. Finally, the experimental box was left in a warm environment for a couple of nights, after which most of the white covering matter had been eaten, leaving only a white stain on the bark. A surprising number (about 50) of undamaged eggs were lying below the twigs.

On the other hand, both adults and larvae of the Pine Ladybird have been observed feeding on the bodies of scale-insects. I have watched one eating a felted scale on an ash trunk, wax covering and all, and leaving only a ring of white fluff on the bark. The idea that an insect can live on wax alone is anathema to the biologist, who insists that protein is needed for development. A tentative conclusion might be that it is attracted to the white waxy protective covering produced by woolly aphids and certain scale-insects, and to such an extent that it only recognises its food when underneath a white protective coat.

A piece of evidence to support this comes from halfway across Europe. In many years of observations of Pine Ladybirds on trunks of ash trees, I have seen only one possible instance of cannibalism. Yet a remarkable photograph by W. Völkl in Klausnitzer (1997) shows its larva attacking and consuming the larva of another ladybird, *Scymnus nigrinus*, which itself has a white waxy protective covering.

Those people who feel that the name of Pine Ladybird has been rendered obsolete, by its colonisation of new habitats, may like to consider the name "shepherd's-pie ladybird" for an insect that likes its protein with a white fluffy covering. Alternatively, if future research were to show that the waxy covering matter is indeed the food, then "cotton-wool ladybird" might be an appropriate name.

## *Hippodamia variegata* (Goeze, 1777) PLATE 9 **Adonis Ladybird**

(= *Adonia variegata*)

**National Status:** Notable (NB)
**Number of Tetrads:** 49
**Status in Surrey:** Very local, probably also migrant
**Habitat:** Low vegetation

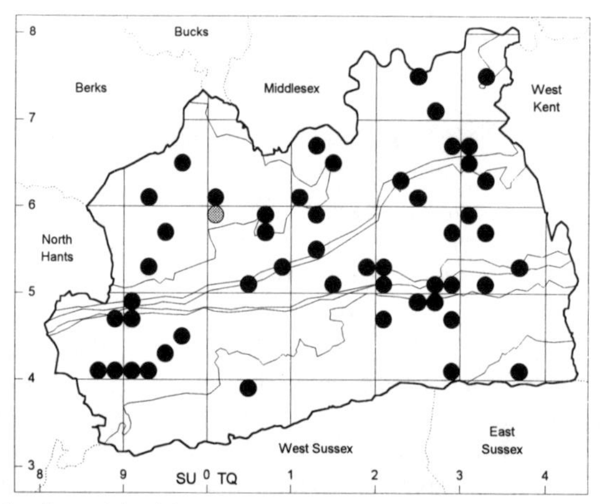


The Adonis Ladybird was given its name by analogy with its scientific name of *Adonia*, but a subsequent change in scientific opinion led to the abandoning of the genus *Adonia* as insufficiently different from *Hippodamia*. The English name need not be subject to such changes but, by popular demand, we have omitted the final apostrophe from Adonis'.

It is a rather small, narrow species superficially resembling the 11-spot Ladybird. The wing-cases appear slightly broader towards the back and have a variable number of black spots on a red background. Examples from Surrey have been reported with three, seven, nine or eleven spots. When few in number, the free spots are concentrated towards the back of the wing-cases, leaving only the combined basal spot with its adjoining white patch at the front.

The pronotum is highly distinctive and offers the easiest way of recognising the species. It is broadest in the middle and has a large central black mark, leaving the front and sides narrowly pale. This mark may have four fingers extending forwards, or the fingers can link up to form a spectacle pattern, or the eyes of the spectacle can disappear leaving a solid black mark. The underside is black with four white spots and the legs are black with brown tibiae and tarsi.

The distinctive larva (not illustrated) is intermediate in pattern between the 2-spot and 7-spot Ladybirds. On the first abdominal segment the two outer tubercles on each side are orange, but on the fourth segment all the tubercles are dark and the central pale mark of the 2-spot is lacking. The early stages have not been recorded during this survey although it is undoubtedly breeding, at least in some years.

This is another highly elusive species. Like the 11-spot, it had a reputation for being a coastal insect so, when a dead example was found in the corner of a Waddon flat on 18 April 1984 (JMcL), it was considered possible that a previous owner had brought it back from the coast accidentally along with the buckets and spades. This impression was short-lived, for a live specimen was then found outside in the garden and two others in a nearby industrial area and on derelict railway land. A preference for urban waste land is borne out by records from elsewhere in the country (e.g. Allen, 1974). There are undoubtedly permanent colonies in the London area, since it has been found repeatedly in a park by the River Wandle on former industrial land (JMcL, DE), while in 1998 many were seen on four separate occasions on derelict land at the mouth of this river (Jones, 1999b),

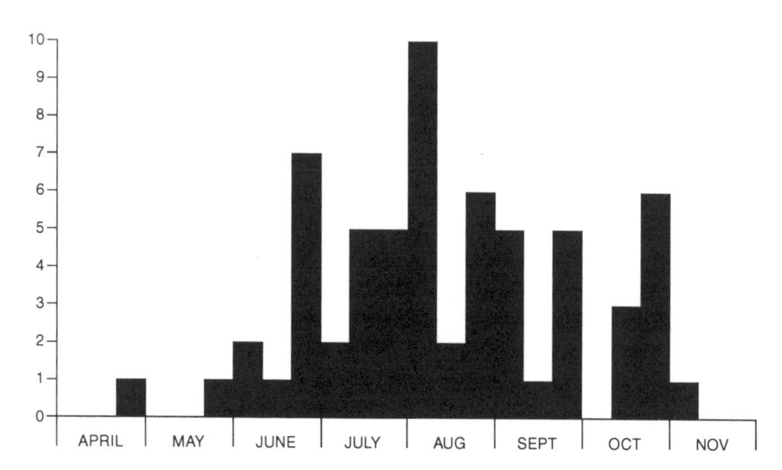


***Adonis Ladybird – number of specimens***

| 1980 | '81 | '82 | '83 | '84 | '85 | '86 | '87 | '88 | '89 | '90 | '91 | '92 | '93 | '94 | '95 | '96 | '97 | '98 | '99 |
|---|---|---|---|---|---|---|---|---|---|---|---|---|---|---|---|---|---|---|---|
| | | ** | | 6 | * | ** | * | | 7 | 11 | 1 * | 4 | 8 | 5 | 10 | 5 * | 4 * | 12 | 13 |

* at least one week of frost in early part of year ** over three weeks of frost

***Adonis Ladybird – number of specimens recorded each year***

Outside the London area the species has been very elusive, occurring mainly as singletons and rarely twice in the same place. It was found with the 11-spot on at least three occasions. Many records in the breeding season refer to aphid-covered weeds of many kinds, but especially goosefoots and oraches, growing on farmland and waste ground. Such habitats are often short-lived, being subject to spraying with pesticides or herbicides, or destruction by cutting or ploughing. This link with a transitory habitat might be one reason for the rarity of the ladybird, but other explanations for this are possible.

This species is very common and widespread around the Mediterranean and I have seen it in great numbers in central Spain, rivalling the 7-spot Ladybird in abundance. This suggests that it is on the edge of its range in Britain and might increase in numbers as global warming takes effect. During the 1990's it does seem to have become more common, with a few now appearing on chalk downland with a partiality for the fruiting heads of wild carrot.

It is remarkable that with so few records the evidence for a partial second generation is better than for the 2-spot and 7-spot Ladybirds which were found in their thousands. It relies on immature adults having an orange background colour. An orange adult on 25 June 1995 represents the first new generation of the year, while a larva (just outside the county) on 5 September is likely to be the second generation. A mating pair with an orange female on 2 September 1996 is early enough for its offspring to survive, while a red example on 20 July 1998 with ill-fitting wing-cases suggests an egg-laden female, but possibly a late survivor of the overwintered generation.

## [*Hippodamia 13-punctata* (Linnaeus, 1758) **13-spot Ladybird**

The 13-spot Ladybird has not been found in Surrey since being listed in the *Victoria County History* (Goss, 1902), apparently on the basis of a record from Battersea published by Fowler (1889). Since that time this location has been used for industry, housing and as a formal park. The last known occurrence of this ladybird in Britain was almost fifty years ago, at Hastings, and it may never have been permanently established in this country, although there is a recent report of its occurrence in Ireland (Majerus, 1994).

This is an elongate species of marshy habitats but it is significantly larger than the Water Ladybird that occurs in similar places. The wing-cases typically have 13 black spots on an orange-red background but the spots may be very faint or may merge. The pronotum is more constant in pattern, with a black rectangular central mark and two isolated spots beside it. Although it is predominantly a wetland species, it has also been found feeding on aphids on field crops in Europe and North America (Hodek, 1973).

The species is well illustrated by Moon (1986) and in Chinery (1986), but the photograph in Majerus (1994, p.182) shows a very similar and closely-related European species, *Hippodamia notata* (Laicharting) (Booth, 1997). The key illustrations of the pronotum in Majerus & Kearns (1989) also appear to show *H. notata*, whilst the painting of the whole insect seems to combine features of both these species.]

## *Aphidecta obliterata* (Linnaeus, 1758) PLATES 11, 14 **Larch Ladybird**

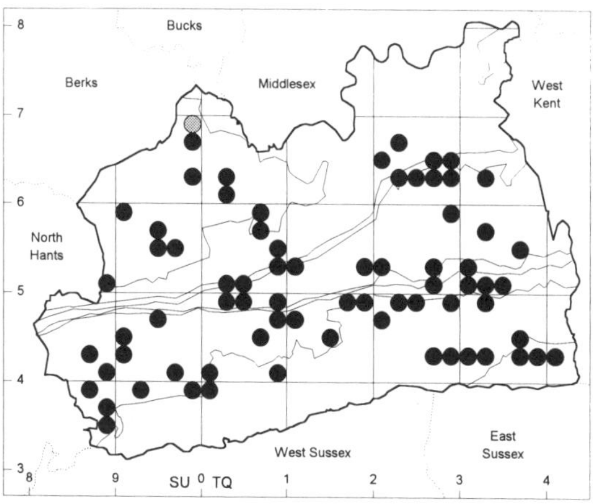


**National Status:** Common
**Number of Tetrads:** 72
**Status in Surrey:** Local
**Habitat:** Restricted to larch, Norway spruce, Douglas fir and occasionally Scots pine

The Larch Ladybird is a confusing species for the beginner since, although it is shaped like a ladybird and behaves like a ladybird, it has no spots. The wing-cases are pale fawn in colour and the chief distinguishing feature is a black mark on the whitish pronotum in the form of a letter 'M'. Although superficially plain, the wing-cases are often speckled with tiny reddish-brown spots. Some specimens have one or two obscure black marks, usually towards the tips of the wing-cases.

In spite of being comparatively featureless, the species is easy to recognise and this is helped by its special habitat. The only possibility of confusion is with those rare examples of the 10-spot that lack spots on the wing-cases and have an M-shaped mark on the pronotum, but the Larch Ladybird is a more elongate species and differs also on its underside, which is much more extensively marked with white.

The larva is a dull grey-brown with most tubercles dark, except for the two outermost on the first abdominal segment which are lemon-yellow and the outer tubercles on the remaining segments, which are white. It is very similar to the larva of the 18-spot and not very different in colour from forms of the 10-spot which might share its habitat. It is best identified using the key in Majerus & Kearns (1989) or reared to adult to confirm its identity.

As its name implies, the Larch Ladybird lives and breeds on larch, but it can also be found on other conifers. Our Surrey records, with numbers of specimens, come from larch (17 adults), Norway spruce (40 adults and 4 larvae), Douglas fir (17 adults and 6 larvae) and Scots pine (20 adults and one pupa). Just three were found on other trees close to one of its regular hosts. There is such a preponderance of Scots pine in Surrey that these figures hide the fact that the ladybird only occasionally occurs on pine, whereas it was found on the other three conifers almost whenever mature trees were examined. We did not search for larvae on larch since the species was already known to breed on this tree at the start of the survey. A record of six adults beaten from an isolated pine tree in a Dorking park on 25 June 1989 suggested that it could also breed on pines and this was confirmed by a pupa and several larvae found on pine in a Horley garden in June 2000. The trees on which the species was found were generally infested with woolly aphids (adelgids) which are known to be its principal food (Majerus, 1994).

Adults were found in all months from April to September, with peak numbers in late August and the first half of September, but the few records of larvae were made in May and June. The only clearly overwintering specimen was one among ivy on the trunk of a Douglas fir on 12 October 1998, but others were beaten from larch in early November and from Norway spruce and Douglas fir in late March, so it seems to overwinter on its host trees (perhaps not on the deciduous larch).

## *Anisosticta 19-punctata* (Linnaeus, 1758) PLATE 9 **Water Ladybird** (= 19-spot Ladybird)

**National Status:** Local
**Number of Tetrads:** 72
**Status in Surrey:** Local
**Habitat:** Margins of ponds and canals

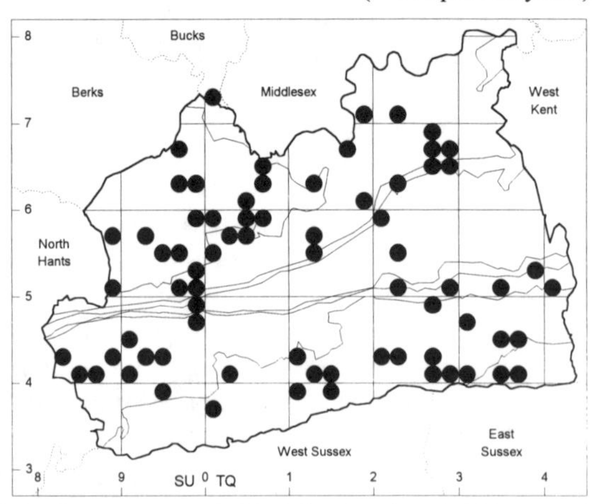


The Water Ladybird has also been called the 19-spot Ladybird (e.g. Moon, 1986) and both names are perfectly appropriate. It is the most well-adapted of our species to living on the water's edge where it feeds on aphids on tall emergent plants such as common reed, reedmace and reed sweet-grass. In winter the ladybirds frequently hide in the narrow gap behind the sheaths of old leaves of reedmace, and many of our records come from finding them in this situation in autumn and early spring.

This is a small (4 mm long) elongate species with densely-packed black spots on its wing-cases and a further six spots on the pronotum. The background colour varies from cream with a pinkish tinge, when the insect is immature in the autumn, to orange-red when it is mature in the spring. Most specimens have nineteen spots on the wing-cases but this number occasionally varies. The pronotum is rounded at the sides and has its greatest width in the middle, which separates this species from all other British ladybirds except the Adonis and the two small *Coccidula* species. Unlike the Adonis, the present species has distinctively pale legs which match the colour of its wing-cases. The underside of the body is black but edged with pale spots.

The larva is plain grey without any contrastingly coloured spots, but its likely identity can often be guessed at from its habitat. For certain identification it needs to be separated on structural details from other drab larvae such as the 16-spot and the small *Coccidula* species which may occur in similar places.

This species breeds rather late in the year. It can be found in its overwintering quarters at least until late April, and we have few records for May. In June it is active and can be seen running up and down the leaves of many species of waterside plants. Mating pairs were observed on 15 June 1994, 21 June 1986, 29 June and 27 July 1997. Larvae and pupae and fresh adults have been found from mid-August onwards, with some slight overlap with the previous generation. By late September many have hidden away for the winter but some are still active in mid-October.

Our photograph shows a larva of the Water Ladybird on a reedmace leaf among a colony of aphids which I believe to be *Hyalopterus pruni*, the mealy plum aphid. This aphid feeds on the leaves of blackthorn and cultivated plums in late spring, then switches to the common reed and other aquatic grasses around midsummer. It is known to be an important prey item for the ladybird (Klausnitzer, 1997) and it is possible that the late breeding season is governed by the presence of colonies of this aphid. Since the ladybird remains on waterside vegetation and does not follow the aphid to the plums, it must feed on other species of aphid when this one is absent. The common reed is not always an easy plant to study when not dressed for wading, but the reeds can be approached and the ladybird seen easily at the Surrey Wildlife Trust's nature reserve at Hedgecourt Pond.

This ladybird can often be found on plants growing well out over the water. On finding a specimen by a pond in 1999, Jonty Denton informed me that he had recently observed that it was a good swimmer. This seemed such a tall story that we decided to put it to an immediate test, and so placed the specimen onto the surface of the pond. At first it fell upside-down, but this seemed unfair so we righted it. It made slow progress with rapid rowing movements of its legs but eventually reached a bent grass stem that dipped into the water at a low angle at both ends. Finding no exit from this stem, the ladybird deliberately launched itself into the water once again and swam to another emergent stem. This was also a dead end but at this point, having proved our theory, we restored the insect to its natural habitat.

The map shows that the Water Ladybird is largely absent from densely built-up areas and from our dry hills of chalk and sand. Elsewhere it is still a local species since it is confined to suitable ponds and other stretches of water. It is common by ponds and along the Wey Navigation but there are no records from faster-flowing stretches of river and very few from the Basingstoke Canal, which is now rather densely wooded. The ladybird seems slow to colonise new ponds and garden ponds. Its requirements for tall emergent plants, left undisturbed in winter and then colonised by aphids (perhaps coming off the plum trees!), seem too demanding for most small gardens.

## *Tytthaspis 16-punctata* (Linnaeus, 1758) PLATES 6, 15 **16-spot Ladybird**

(= *Micraspis 16-punctata*)

**National Status:** Local
**Number of Tetrads:** 333
**Status in Surrey:** Almost ubiquitous
**Habitat:** Grassland

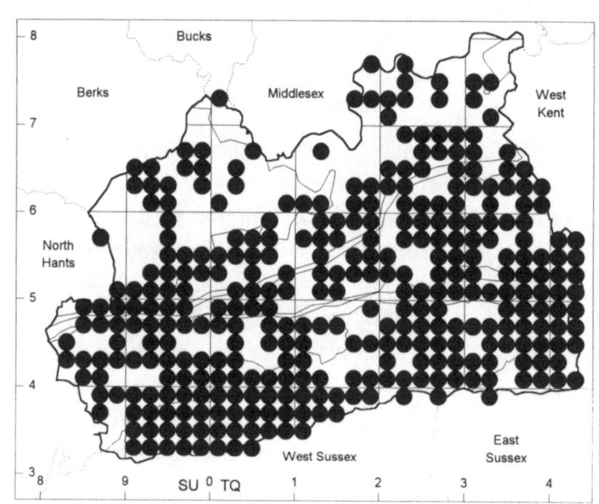


The engaging little 16-spot Ladybird is our most abundant species, but even so it is not at all well-known, being small (2.5 to 3 mm long) and living close to the ground. It is usually found in grassland where it occurs in huge numbers at some sites but is entirely absent from others. It is a constant companion on picnics in the Surrey countryside, and the observant picnicker will soon notice it running frantically up and down grass stems and even investigating the lunch-box. But what is it looking for? This appears to be something of a mystery, for it never seems to stop.

The background colour is pale cream and the 16 spots of its name are arranged in a central row of three and an outer row of five on each wing-case. When seen from above, the two inner rows of three spots resemble a six of dice. The three central spots of the outer row are usually linked into an irregular line. The only variation found in Surrey is for this line to extend and include another spot, or to be shorter and leave an additional spot free. I have seen but one example with all the spots free, while one remarkable specimen had the usual pattern on one wing-case but all the spots free on the other. Among thousands of specimens examined, one or two have been a little bit orangey in colour, but none as distinctly flushed with red as reported and illustrated by Jones (1999a).

An additional feature of the 16-spot is a narrow black line of uniform width along the inner edge of the wing-cases. Outside this and parallel to it is a line of translucent dots. The pronotum has two black spots outside a semicircle of four black spots with no central spot.

The larva is short-spined, about 5 mm long, and plain dark grey with a paler side-stripe. It can be identified easily under a lens by examining a structural detail of its thorax (an additional hairy tubercle on the sides of the second and third segments), but it is not so easy to name at a glance. Because of this, we have few records of 16-spot larvae, although they are undoubtedly just as abundant in their season as the adults are in autumn. We are thus unable to offer any insight into the feeding habits of the larvae and the number of generations a year. These are questions of particular interest in this species, which is known to eat different types of food and increases spectacularly in numbers in the autumn. The few records of larvae were made on 1 September 1984 on a rye-grass head, 26 August

1989, 12, 20 & 24 July 1995, 29 June and 11 August 1997, and 4 & 19 July 1998. Mating pairs were observed on 18 May and 4 July 1997.

The food of the 16-spot Ladybird is a somewhat controversial matter. It was long thought to prey on aphids and is indeed classified, by reason of its mouth-parts, in a subfamily of the Coccinellidae whose members generally feed upon aphids. More recent studies have found that it can feed upon mildew, like the 22-spot and Orange Ladybirds, but these species are regularly and consistently observed feeding on mildew, both as larvae and as adults, while records of the more abundant 16-spot feeding on mildew are few. A recent publication lists its foods as mildew, thrips, mites and the pollen of grasses (Klausnitzer, 1997). The problem is to determine its staple food, if there is one, by examining what the larvae are eating in the wild.

Our Surrey records of feeding illustrate the problem rather than help to solve it. On 11 August 1997, at Woodlands Park near Leatherhead, about six 16-spots were on the flowers of scentless mayweed and clearly nibbling at the disc florets, while ten more were on the mildewed leaves of meadow buttercup and knotgrass. If a backwards-and-forwards motion indicates feeding, then they were eating white mildew on the leaves of these plants. This motion was also observed on a dock leaf on 30 June 1985 at Haxted. Just outside the county at Spring Park in West Kent, 130 ladybirds were seen on dandelion flowers on 23 May 1996, suggesting feeding on pollen. In 1984 two were seen on mildewed hogweed leaves in the company of 22-spots, while two more were on the mildewed leaves of young oaks favoured by the other species. These records indicate feeding on mildew and pollen but are so few that both might only be occasional foods rather than regular ones.

This species is well-known for its habit of overwintering in clusters in exposed situations. Two such instances were recorded in Surrey in the winter of 1984/85. On 10 November 1984 I found 3300 on the trunks of a few ash trees beside a stream in a small valley at Hound House Farm near Newdigate. They were in clusters of up to 100 in depressions and crevices in the bark, from near ground level to 5 metres above ground, and mostly on the north and west side of the trees, away from the stream and beside some rather damp, sloping pasture. A month later a friend and neighbour, John Steer, was planning a country walk and I recommended that he visit this site. He chose to go in quite the opposite direction and returned with a tale of uncountable numbers, thousands and thousands, on the posts of a barbed-wire fence along the bridleway between Horley and Smallfield. The word 'uncountable' presented a challenge to one trained in mathematics, so next day I went out and counted them, using a technique designed for counting flocks of birds: first counting the individuals in a small area marked out by part of one's hand, then measuring the number of such areas in the whole assembly. The resulting total over 100 posts along 200 metres of fence was 17,100 ladybirds; the date was 9 December 1984. The clusters were on concrete posts as well as wooden ones, and also on barbed wire and dead stems of grass, nettles and other plants. An unusual cause of mortality had arisen from a horse leaning against a post, squashing many ladybirds. On returning to the site on 4 May 1985, no ladybirds remained exposed but many were found under dead grass, in leaf litter and around worm casts at the foot of the posts. In the following winter the bark was peeling from the wooden posts and most exposed clusters were on the concrete posts, totalling 560 ladybirds on 1 December 1985. More were in dead vegetation and in leaf litter around the base of the posts. On 31 March 1986 there were about 500 at the bases of the posts (JBS).

At first it seemed possible that ladybirds came from far and wide to join these overwintering clusters, but the high numbers found in grass tufts beside fields at many sites in Surrey have convinced me that this is not the case. The pasture beside the bridleway is four hectares in area, so spreading the overwintering population over the whole field would result in one ladybird to every two square metres, a density often exceeded when examining grassland in summer.

Our total of 17,100 may seem a small thing compared with the "million-ladybird field" reported from Kent by Richard Jones (1999a), but it is a direct count and so a minimum figure, while the million is an estimate which may err in either direction. Even if exaggerated, this population would still be several hundred thousand. The field is clearly a large one. One million ladybirds spread over 20 hectares would give an average density of five per square metre, which is by no means impossible.

Some explanation is needed as to why these exposed clusters occur only in certain places and in certain years. I suggest that local flooding may be the cause by which ladybirds overwintering in tufts of long grass around fence-posts are forced up the posts by the rising water level, but have as yet no direct evidence of this. The Newdigate site was beside a stream and the Horley bridleway was at that time frequently flooded in winter and passable only in wellingtons, although it has since been raised up, surfaced with rubble and drained by a deep ditch. On 23 April 2000 there were no ladybirds on the posts but some were found beneath dead grass and leaf litter wherever this was searched.

Only two overwintering clusters have been reported since the two in 1984/85. There were several hundred, probably thousands, on an oak tree at Chilworth on 31 December 1989. Most were in crevices in the bark and there were also a few on a nearby fence (NAC). In the late autumn of 1995 there were about 600 on a stile near Dorking. Just 27 returned to this stile in September 1996 and all had gone before the end of the year (GLAC).

The majority of records on the distribution map were made in autumn by tapping tufts of grass or other plants sideways over an entomologist's beating tray. This has shown that large numbers are present at very many sites in Surrey. Finding 50 to 100 in a single tap is commonplace, while on two occasions 400 ladybirds were knocked from single plants of nettle and creeping thistle.

In spite of being so abundant, it does not occur absolutely everywhere in the county, being uncommon on heathland and largely absent from built-up areas, although occasionally found in gardens and allotments (JMcL, MDB). It is not even found in every grassland. My most reliable indicator for finding it is the presence of grazing animals. It was in great abundance at Horne in 1998 at the edges of permanent pasture, organically farmed and close-grazed by sheep. There were no flowers here among the grass and their presence does not seem to be necessary. It is usually absent from improved pastures that have been ploughed and re-sown with rye-grass. Of course the use of insecticides may kill off a population or even, if the ladybird is dependent on mildew, the use of fungicides.

Grazing is not absolutely necessary, for it is found in areas of long grass such as those parts of Wandsworth Common where the grass is left uncut. Regular mowing seems to be detrimental but a few are occasionally found around the edges of cricket fields, and it was numerous in the few clumps of long grass that had escaped the mower at Brockwell Park in 1999. It is however often absent from areas of long grass, even where other species are common. It is frequently found walking over very short turf or on bare ground among grass, and in one instance on leaf litter on a river bank under trees with very little live vegetation at all (26 May 1997, Chiddingfold). It was absent from apparently suitable horse-grazed pasture by the River Wey at Shalford in 1998, but present on rising ground to the west. It can also occur on farmland in small pieces of suitable habitat such as the grassy headlands around arable fields. Very occasionally it climbs up on trees or bushes – one was beaten from Douglas fir and another from ivy on an oak trunk.

The opinion of many authors (Fowler, 1889; Joy, 1932; Duffy, 1945; Pope, 1953), that the 16-spot Ladybird favours marshy areas, is not borne out by this survey, nor is a preference for sandy places (Fürsch, 1967). It is found generally in Surrey, on dry hills of chalk or sand as well as in low-lying fields on clay soil.

## *Adalia bipunctata* (Linnaeus, 1758) PLATES 2, 3 **2-spot Ladybird**

**National Status:** Common
**Number of Tetrads:** 367
**Status in Surrey:** Ubiquitous in towns, local elsewhere
**Habitat:** Broad-leaved and coniferous trees, and tall herbaceous vegetation

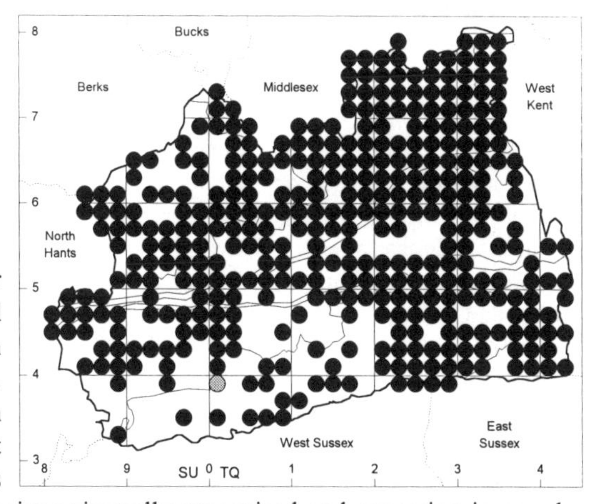


The 2-spot is the City Slicker among the ladybirds, being found in much greater numbers in urban areas than in the countryside. Owing to a liking for aphids on roses and other garden plants, it might also be called the Gardener's Friend. It is one of the two species universally recognised and appearing in popular portrayals of ladybirds. It forces itself on our notice by hibernating on or around our houses and coming indoors in the spring.

The pattern of black and red shows a great deal of variation, but the typical form with a large round black spot on each wing-case is vastly more common than the melanic form with either four or six red spots on a black background. Even less common are intermediate forms which show a continuous range of variation with the round spot becoming either a comma or a band, with or without an additional loop behind it. Further extension of the black mark gives a chequered form with equal amounts of red and black, but this variety is rare in this species, as is the extreme melanic with two broad shoulder spots (these two forms are more common in the 10-spot Ladybird). Melanic 2-spots may be either male or female and are often seen mating with typical ladybirds, or occasionally with other melanics. During this survey 95% were recorded as typical, 4% as melanic with 4 or 6 red spots, and 1% as intermediates with comma-shaped spots, band, or band with loop (1200 specimens were checked). This overall figure for the proportion of melanics masks considerable local variation. There are noticeably more melanic 2-spots towards the north-west of the county. Together with adjoining areas in neighbouring counties, this forms a 'hot-spot' for melanism in the 2-spot Ladybird with the proportion of melanics about 20% (John Muggleton, *pers. comm.*).

The underside and legs are completely black, except in freshly-emerged specimens, and this distinguishes the 2-spot from similar varieties of the 10-spot Ladybird. The four-spotted melanic form may look very similar to a Pine Ladybird and occur in similar places, but the wing-cases of the 2-spot slope evenly and lack the turned-up rim of the other species.

The larva is normally dark grey with three orange spots forming an obvious triangle when viewed from above. These spots are in the centre of the fourth abdominal segment and on the middle tubercle on each side of the first segment (often including the outer tubercle as well). The dark outer tubercles on segments 5 to 8 separate it from the 10-spot Ladybird. Unfortunately, the pattern is rather variable. In one variation, the outer tubercle of segment 4 may also be orange. Alternatively, some or all of the orange spots may be absent and the larva may then be confused with other, rarer species such as the Adonis Ladybird.

During this survey 2-spots have been seen among or feeding on a variety of aphids on a large number of host-plants. Among the most important are nettles and tall umbellifers such as hogweed and cow parsley, all of which play host to well-camouflaged green aphids. The ladybird will prey on aphids on garden plants such as roses, sunflower and various ornamental shrubs, wherever these are left unsprayed. It feeds on brown aphids on tansy and attacks the black aphids that cause cherry leaves to curl and those that infest elder bushes and various low plants such as goosegrass and thistles. The ladybird is common on birch and sallow and sometimes found far from houses on these trees. Conifers are not important hosts for this species but it is occasionally found on Scots pine and has been seen eating grey aphids on Lawson's cypress. Many of these aphids are among the large numbers of prey items listed by Mills (1981). Some unusual foods were recorded in the present survey: one nibbling the stigma of a flower of a box tree on 23 April 1984 and another eating a yellow beetle larva on a sallow catkin on 23 May 1984.

The link with towns is demonstrated by comparing the distribution map for the ladybird with the map of urban areas. There is a particular contrast between our records for the London area (TQ17/27/37), with the 2-spot everywhere, and for the attractive countryside north-east of Haslemere (SU93) where the two dots pick out the villages of Wormley and Chiddingfold. When numbers are considered, the urban link becomes even stronger. The 2-spot made up almost one-half (48%) of all the ladybirds seen around Kew (TQ1977) between 1963 and 1968 (Eastop & Pope, 1969). My own records from the present survey list just two 2-spots from SU93 among 816 ladybirds of 24 different species (0.25%). The reason for this link with urban areas is difficult to find, but the 2-spot is the only one of our common species that regularly hibernates around buildings. The obvious explanation is

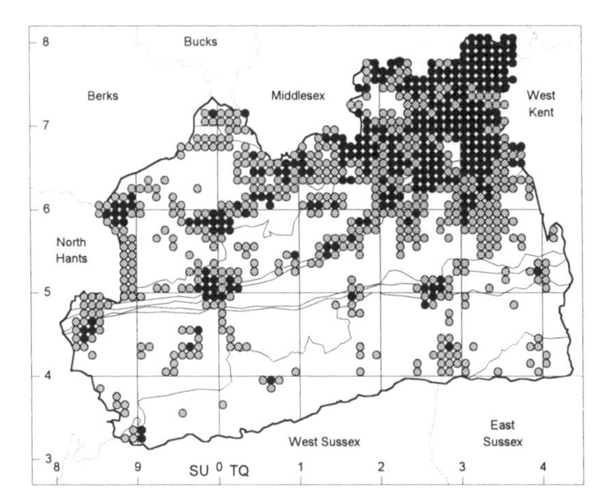


***Major urban areas in Surrey***

that the warmth from buildings enables the ladybirds to survive periods of extreme cold in winter, but this is difficult to sustain when considering that the 2-spot is a common species across Central and Eastern Europe where the winters are far colder than ours.

In Surrey, overwintering in or around buildings has been observed in cities (London), towns (Horley) and villages (Clark's Green) and is probably a universal habit. Industrial buildings and office blocks seem particularly favoured. In autumn 2-spots can also be seen investigating cracks in trees and fence-posts, and they can certainly survive some winters in such sites.

In 1984 I resolved to investigate the number of generations produced by this species in a year, and was greatly surprised by my findings. Observations could be made on a daily basis around the streets of Horley before and after working in an office. There was a large population which first appeared in numbers during the first spell of fine weather in late April. For much of May the weather was mainly dull and cool, although dry, and the ladybirds appeared somewhat reluctant to start the breeding cycle, being more intent on feeding. The weather remained cool but became unsettled with much rain towards the end of the month when many 2-spots were seen mating: 21 pairs on 30/31 May. Further mating pairs were seen regularly throughout June and into July but in declining numbers. Larvae and pupae were observed in June and the first fresh adults on 10 July, recognisable by their orange background colour. From then on these new adults appeared in ever greater numbers. In mid-July overwintered adults were still present and still mating, but the last such pair was seen on 22 July, by which time the new generation was becoming numerous, with groups of 40 to 100 noted on four occasions. Here came the surprise. From this date onwards, although some of the new ladybirds later took on a red colour, no further mating pairs were seen. It is thus fairly certain that there was only a single generation of 2-spot Ladybirds in 1984, at least in south-east Surrey. This was a year of a late spring, in which all natural phenomena were very late, and of hot dry weather in July when the new generation of ladybirds came through (source of weather notes: *Nature in Cambridgeshire*, no. 27).

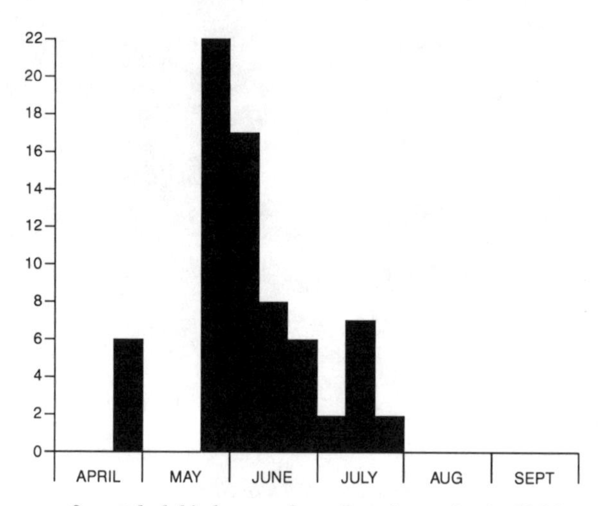


***2-spot ladybird – number of mating pairs in 1984***

Observations in the year 1985 followed much the same pattern as the above, with the overwintered generation in decline but overlapping with the new generation when this was first seen on 14 July. The only three mating pairs observed on or after this date had all overwintered. Recording of this species was much less intensive in subsequent years, but significant variations from the pattern were noticed. Mating activity began much earlier in some mild springs, even in March in 1990 and 1995. Two orange-background ladybirds were seen mating on 11 July 1986 and two more such pairs on 13 June 1992. Mating pairs of unspecified colour on 7 August 1996 and 2 August 1997 were probably of the new generation, especially since many pairs had been seen in early April in the latter year. Finally, two freshly-moulted adults were found in Dulwich on 14 October 1999. These few observations offer clear evidence that a small, partial second generation is produced in some years.

## *Adalia 10-punctata* (Linnaeus, 1758) PLATES 1, 3 **10-spot Ladybird**

**National Status:** Common
**Number of Tetrads:** 301
**Status in Surrey:** Ubiquitous
**Habitat:** Broad-leaved and coniferous trees

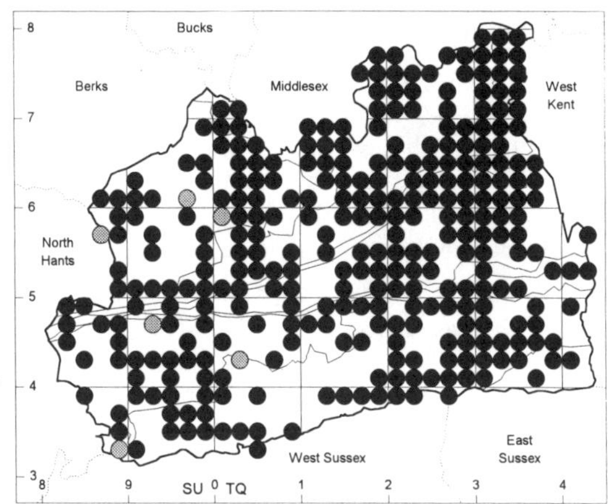


This small ladybird is common but not often seen unless searched for, since it spends most of its time among the foliage of trees feeding upon aphids. It is one of the more variable species but the problem of identification can be simplified by grouping the varieties into three distinct types. The typical form has ten small black spots on the wing-cases on a background that varies from greenish-yellow in fresh specimens to orange-red in mature individuals. The spots can be small and faint or strong and dark, and their number varies, usually between six and twelve but sometimes there are none at all. Although the spots on the wing-cases are variable, the pattern on the pronotum is constant in the typical form, with four black spots in a semicircle around a basal spot on a whitish background.

The second distinct variety is the chequered form in which the spots have linked up so that there are equal amounts of red and black, or brown and yellow in fresh examples. The third variety is the melanic form in which the wing-cases have a comma-shaped red shoulder-stripe on a black background. Since the spots on the pronotum tend to join up in the chequered form and merge completely in the melanic form, these varieties are best identified from their undersides. In all 10-spot Ladybirds the underside is mostly orange-brown and even on the darkest specimens some brown colour remains, at least on the legs. A narrow white triangle is always present outside the base of the middle legs, but is generally obscured when the insect is at rest.

These underside features will always separate 10-spots from similar varieties of the 2-spot Ladybird. During this survey I have had difficulty in distinguishing only two specimens, and in such cases the possibility of a hybrid must be considered. The two species are closely related and capable of interbreeding, although the resulting offspring are infertile (Majerus, 1994). This has been demonstrated using captive specimens and it is doubtful whether hybrids ever occur in the wild.

The 10-spot can be superficially similar in colour and pattern to the much larger Cream-streaked Ladybird with which it can occur on conifers. In rare instances, extra spots occur on the pronotum so that it mimics its larger relative closely, although the number and position of spots on the wing-cases are always different and the underside of the Cream-streaked has much more white on it. I have also twice found 10-spots that were perfect mimics of a Larch Ladybird and occurred in one of its habitats, on spruce trees. The wing-cases were spotless and the spots on the pronotum had joined up to form an 'M'. The shape was wrong, for the Larch Ladybird is more elongate, so I checked the undersides to confirm their identity.

The larva closely resembles that of the 2-spot, with a triangle of yellow spots formed by a patch in the centre of the fourth abdominal segment and the middle tubercle on each side of the first. The background colour of the 10-spot larva is pale grey (dark grey in 2-spot) and the outer tubercles on abdominal segments 5 to 8 are all whitish (dark in 2-spot). Comparatively few larvae have been found in this survey, probably because they are usually out of reach up the trees, and these were reared out to ensure correct identification.

Exactly 10% of the 590 specimens whose habitat was noted were on herbaceous plants, while 90% were on trees and shrubs. Many of the former were on nettles (5.6%), especially in the spring. A mating pair was found on white dead-nettle and two freshly-moulted adults on nettles but, since the specimens on herbaceous plants were generally found beneath trees, I prefer to interpret them as having dropped off the trees rather than showing any tendency to breed on herbaceous plants.

Although the main habitat of the 10-spot is broad-leaved trees, it can also be found (23% of specimens) on conifers such as Scots pine and Norway spruce, against whose buds it is beautifully camouflaged. Larvae and fresh adults were found on both these trees in small numbers, and also on Douglas fir. Some records from Lawson's cypress were mostly from early or late in the year, suggesting overwintering among the foliage, but include a mating pair and a fresh adult.

More usual habitats, on which the 10-spot occurs in quantity, are common oak, lime, sycamore, hawthorn, sallow and birch, on all of which either fresh adults or larvae were found. Further records refer to many other broad-leaved trees and shrubs. In spring, before the leaves open, 10-spots can be found on tree trunks in urban areas but in much smaller numbers than the 2-spot. The species is often found on hawthorn just as the buds are bursting into leaf. Beating hawthorn at this time is the most reliable way of checking its presence in an area, although it may have disappeared completely from the hawthorn by the time the flowers open in May. Most larvae seem to pupate on or near the leaves where they have been feeding, but a few pupae were found in July in cracks in the bark of lime trunks, even before the advent of *Pulvinaria regalis* scale-insects. Our photograph (*Plate 3*) shows a larva, pupae and freshly-emerged adults close by the egg-masses of *Pulvinaria*, but we have no direct observations of feeding on this scale-insect (DE).

The 10-spot Ladybird probably occurs everywhere in Surrey. It is common in urban areas on trees at roadsides, in gardens and in parks. It is also found fairly generally on the trees and tall hedges which abound in the Surrey countryside. There are, however, blanks on the distribution map which I take to be locations surveyed at times when the species was difficult to find. For example, an attempt to fill the gap between Cranleigh and Ewhurst (TQ0840) produced no 10-spots from much suitable habitat on 1 May 2000. There is no reason why the species should be absent from this area and I assume that either the overwintered adults had died off through bad weather, or the survivors had moved out of reach up the trees. Ten other species of ladybird were found that day, but all in small numbers.

The introduced, evergreen, holm oak can be infested with pale green aphids among which the 10-spot often occurs in numbers and is certainly breeding. The ladybird has also been found among the foliage of holm oak in winter (AJW). Other overwintering sites appear to be in leaf litter, in bark crevices on tree trunks, and among the leaves of ivy or Lawson's cypress.

Whenever a 10-spot Ladybird was encountered during the survey, its variety was noted, at least for the majority of specimens. Out of a total of 584, 70% were typical, 22% chequered and 8% melanic. There was little change in these ratios throughout the survey and they are in broad correspondence with similar counts made previously in Surrey and elsewhere in Britain (Merritt Hawkes, 1927; Eastop & Pope, 1966; Majerus, 1994).

| | | 1984 | '85 | '86 | '88 | '89 | '90 | '93 | '94 | '95 | '97 | '98 |
|---|---|---|---|---|---|---|---|---|---|---|---|---|
| **June** | 1st-10th | | | | | | | | | | | |
| | 11th-20th | | | | | | | | | | | |
| | 21st-30th | | | | | 6 | 2 | | | | 4 | |
| **July** | 1st-10th | | | | 4 | | 1 | | 2 | | | |
| | 11th-20th | 7 | 1 | 1 | 13 | | 1 | 1 | 1 | | | |
| | 21st-31st | 2 | | 3 | 1 | | | | 2 | 3 | | |
| **Aug** | 1st-10th | | | | | | | | | 2 | | 1 |
| | 11th-20th | | | | 1 | | | | | 3 | 1 | 3 |
| | 21st-31st | | | | | 3 | 2 | | 3 | 2 | | 13 |
| **Sept** | 1st-10th | | | | 6 | | | 2 | 1 | | | 3 |
| | 11th-20th | | | | | | | | | | | 5 |
| | 21st-30th | | | | | | | | | | | |
| **Oct** | 1st-10th | | | | | | | 1 | | | | 1 |
| | 11th-20th | | | | | | | | | | | 1 |
| | 21st-31st | | | | | | | | | | | |

***10-spot ladybird – dates of freshly-emerged adults***

The evidence that this species produces a second generation late in the year is strong but not totally conclusive. In most years there seem to be two distinct peaks of adult emergence about two months apart, either in late June and late August or in the first half of July and early September. Occasional fresh specimens are found between these peaks and the possibility of a single generation with a protracted emergence, corresponding to an extended period of egg-laying in the spring interrupted by spells of bad weather, cannot be discounted. Only ten mating pairs were reported during the survey, dating from April to early June with one exception. On 24 June 1989, at a time when several empty pupal cases and freshly-emerged adults were being found, a male with a red background was mating with a female with an orange background on a small oak tree in Horley. Three days later a 10-

spot was laying eggs on the leaves of a lime tree in Charlwood. These examples may however be late survivors of the overwintered generation.

Dates of fresh-looking adults are given in the table, which relies on a subjective judgement of the background colour of the adult ladybird in its first few days of life, but those distinct peaks observed in 1986, 1990 and 1998 include larvae or pupae that were reared out. The table also needs to be approached with some caution since recording was variable over the years. Provisionally, in the years 1984-6 there was a single clear emergence in mid-July, but in all the years 1988-90, 1993-5 and 1997-8 there was some evidence for two distinct peaks of emergence. A larva on 23 September 1998 (reared to adult by 8 October) and another on 15 October 1999, very probably this species but not certainly determined, may well represent a third generation.

## *Coccinella 7-punctata* Linnaeus, 1758 PLATES 3, 4, 5 **7-spot Ladybird**

**National Status:** Common
**Number of Tetrads:** 517
**Status in Surrey:** Ubiquitous
**Habitat:** Low-growing plants of many kinds

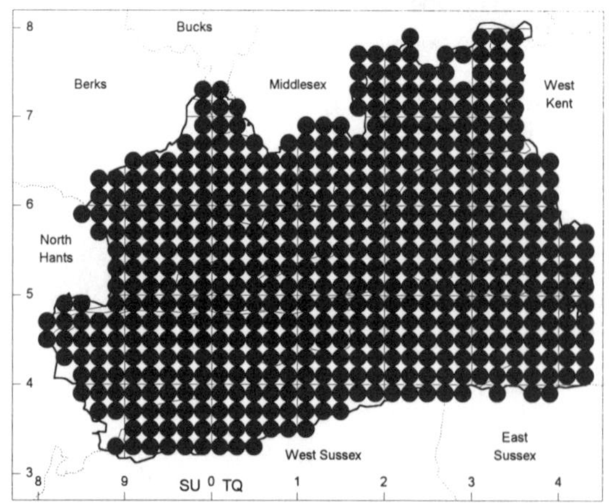


The 7-spot is the ladybird of popular imagination and is very easy to recognise. The pattern of seven black spots on a red background is virtually constant, the only slight variation being in the size of the spots. It is our most widespread species and sometimes also the most abundant, but its numbers fluctuate greatly from year to year. With its large size, bold warning colours and habit of sitting exposed on low herbaceous vegetation, it is difficult to overlook.

It can only be confused with the Scarce 7-spot, a rare species that is only found in the immediate neighbourhood of the nest mounds of wood ants. If the spot in the centre of the wing-case is larger than the one at the sides, then the 7-spot can look somewhat like its rarer relative. In these doubtful cases, its identity can be confirmed by looking at the underside where the 7-spot only has a single pair of white spots, outside the base of the middle legs. There is little need to check this underside feature on typical ladybirds with the three free spots on each wing-case of roughly uniform size.

Although the adult 7-spot is so well-known, its larva goes almost completely unrecognised by the general public and even among entomologists. In its years of abundance it can be found everywhere on roadsides, in gardens and on almost every patch of herbaceous vegetation. To know this creature, just one fact needs to be remembered – it has eight spots! The full-grown larva is blue-grey in colour and the orange spots are arranged in two

DAVID ELEMENT

*Adalia 10-punctata* – freshly emerged

GRAHAM COLLINS

*Adalia 10-punctata* – immature typical

DAVID ELEMENT

*Adalia 10-punctata* – mature typical

DAVID ELEMENT

*Adalia 10-punctata* – strongly- and faintly-spotted typical, mating

DAVID ELEMENT

*Adalia 10-punctata* – chequered

ANDY CALLOW

*Adalia 10-punctata* – melanic

***10-spot Ladybird – varieties***

GRAHAM COLLINS

*Adalia bipunctata* – mating pair, typical

DAVID ELEMENT

*Adalia bipunctata* – enlarged spot

DAVID ELEMENT

*Adalia bipunctata* – banded

GRAHAM COLLINS

*Adalia bipunctata* – band and loop

ANDY CALLOW

*Adalia bipunctata* – six-spot melanic

DAVID ELEMENT

*Adalia bipunctata* – four-spot melanic

***2-spot Ladybird – varieties***

ANDY CALLOW

*Adalia bipunctata* – 2-spot laying eggs

RICHARD JONES

*Adalia bipunctata* – larva feeding on pre-pupa

DAVID ELEMENT

*Adalia bipunctata* – 2-spot pupae

DAVID ELEMENT

*Adalia 10-* & *bipunctata* – larva, pupae, fresh adults and scale insects

ANDY CALLOW

*Coccinella 7-punctata* – 7-spot larva eating aphids

DAVID ELEMENT

*Coccinella 7-punctata* – 7-spot pupa

***Ladybird development***

GRAHAM COLLINS

*Formica rufa* – nest of wood ants

GRAHAM COLLINS

*Coccinella magnifica* – Scarce 7-spot

GRAHAM COLLINS

*Coccinella magnifica* – Scarce 7-spot

GRAHAM COLLINS

*Coccinella 7-punctata* – 7-spot

DAVID ELEMENT

*Coccinella 7-punctata* – 7-spot, immature

DAVID ELEMENT

*Coccinella 7-punctata* – 7-spot, freshly emerged

***Common and Scarce 7-spots***

DAVID ELEMENT

*Coccinella 7-punctata* – taking off

TONY MUNDELL

*Coccinella 7-punctata* – emerging from hibernation

ANDY CALLOW

*Coccinella 7-punctata* – attacked by ant

ANDY CALLOW

*Coccinella 7-punctata* – beneath yellow umbel

ANDY CALLOW

*Coccinella 7-punctata* – aphid's-eye view

GRAHAM COLLINS

*Coccinella 7-punctata* – parasitised

***7-spot behaviour***

DAVID ELEMENT

*Propylea 14-punctata* – 14-spot, freshly emerged

RICHARD JONES

*Propylea 14-punctata* – 14-spot, mating pair

GRAHAM COLLINS

*Propylea 14-punctata* – 14-spot, spots free

DAVID ELEMENT

*Propylea 14-punctata* – 14-spot, spots joined

DAVID ELEMENT

*Tytthaspis 16-punctata* – 16-spot

TONY MUNDELL

*Tytthaspis 16-punctata* – 16-spot, overwintering cluster

***14-spot and 16-spot Ladybirds***

DAVID ELEMENT

*Psyllobora 22-punctata* – 22-spot, mating pair

DAVID ELEMENT

*Psyllobora 22-punctata* – 22-spot, hibernating

ANDY CALLOW

*Halyzia 16-guttata* – Orange Ladybird

GRAHAM COLLINS

*Psyllobora 22-punctata* – 22-spot, grazing mildew

GRAHAM COLLINS

*Calvia 14-guttata* – Cream-spot, immature

DAVID ELEMENT

*Calvia 14-guttata* – Cream-spot

***Mildew-feeding ladybirds, and Cream-spot***

RICHARD JONES

*Epilachna argus* – Bryony Ladybird

GRAHAM COLLINS

*Epilachna argus* – Bryony Ladybird, larva

GRAHAM COLLINS

*Subcoccinella 24-punctata* – 24-spot

ROBERT POPE

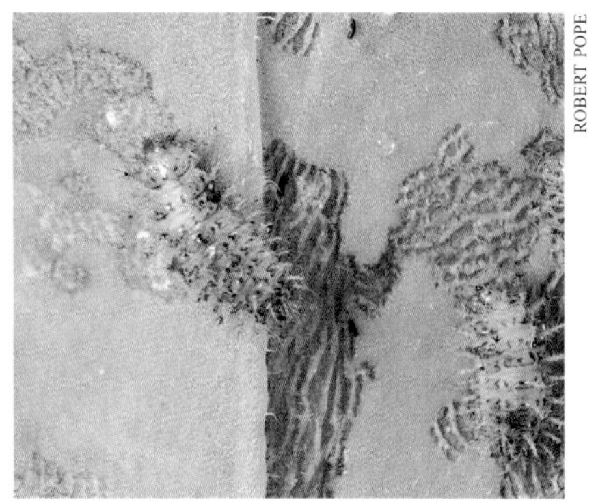

*Subcoccinella 24-punctata* – 24-spot, larvae feeding

DAVID ELEMENT

*Subcoccinella 24-punctata* – 24-spot, mating pair

DAVID ELEMENT

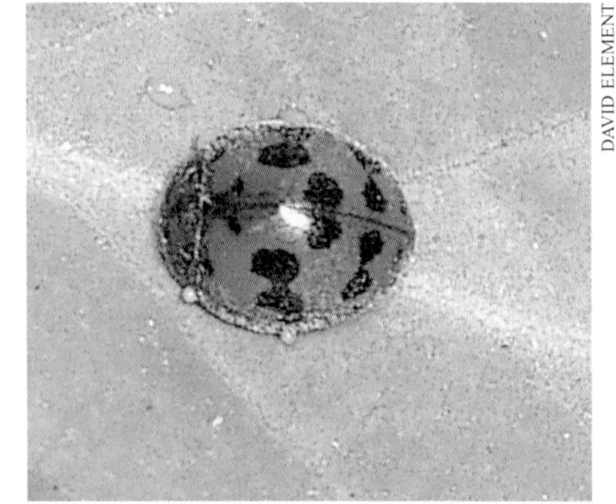

*Subcoccinella 24-punctata* – 24-spot, reflex bleeding

***Plant-feeding ladybirds***

GRAHAM COLLINS

*Hippodamia variegata* – Adonis' Ladybird

DAVID ELEMENT

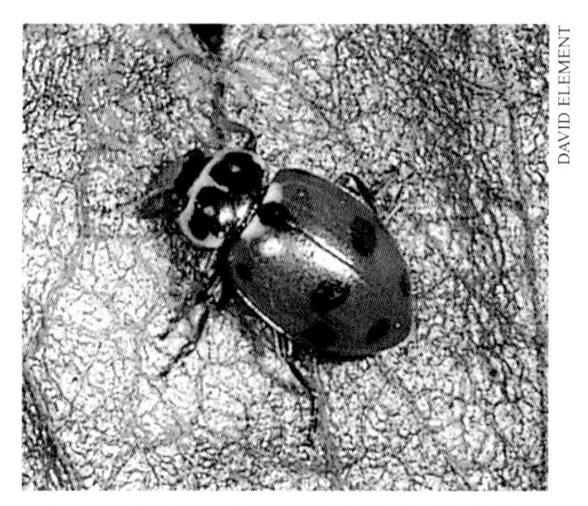

*Hippodamia variegata* – Adonis' Ladybird

DAVID ELEMENT

*Hippodamia variegata* – Adonis' Ladybird, underside

GRAHAM COLLINS

*Anisosticta 19-punctata* – Water Ladybird, larva

DAVID ELEMENT

*Anisosticta 19-punctata* – Water Ladybird, autumn

GRAHAM COLLINS

*Anisosticta 19-punctata* – Water Ladybird, winter

***Adonis' and Water Ladybirds***

DAVID ELEMENT

*Harmonia 4-punctata* – Cream-streaked Ladybird

GRAHAM COLLINS

*Harmonia 4-punctata* – Cream-streaked Ladybird

DAVID ELEMENT

*Harmonia 4-punctata* – Cream-streaked Ladybird, underside

DAVID ELEMENT

*Myrrha 18-guttata* – 18-spot, underside

DAVID ELEMENT

*Myrrha 18-guttata* – 18-spot

GRAHAM COLLINS

*Myrrha 18-guttata* – 18-spot

***Ladybirds of pine trees***

GRAHAM COLLINS

*Anatis ocellata* – Eyed Ladybird

DAVID ELEMENT

*Anatis ocellata* – Eyed Ladybird

JIM PORTER

*Myzia oblongoguttata* – Striped Ladybird

DAVID ELEMENT

*Aphidecta obliterata* – Larch Ladybird

GRAHAM COLLINS

*Chilocorus bipustulatus* – Heather Ladybird

JIM PORTER

*Coccinella hieroglyphica* – Hieroglyphic Ladybird

***Ladybirds of conifers and heathland***

GRAHAM COLLINS

*Chilocorus renipustulatus* – Kidney-spot Ladybird

TONY MUNDELL

*Chilocorus renipustulatus* – Kidney-spot, cluster of pupae

DAVID ELEMENT

*Exochomus quadripustulatus* – Pine Ladybird, with eggs of *Pulvinaria* scales

DAVID ELEMENT

*Exochomus quadripustulatus* – Pine Ladybird, immature

GRAHAM COLLINS

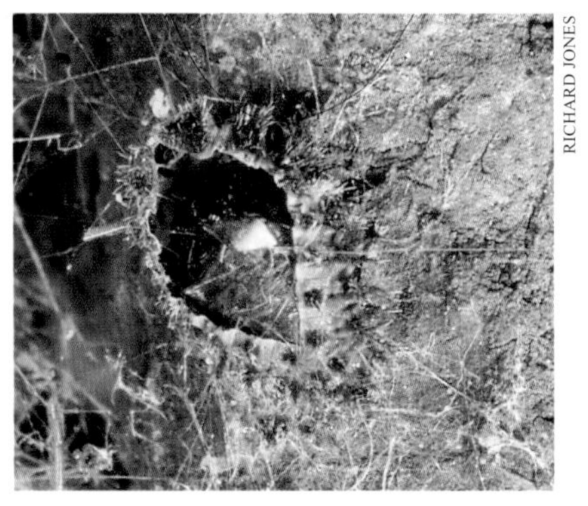

*Exochomus quadripustulatus* – Pine Ladybird

RICHARD JONES

*Exochomus quadripustulatus* – Pine Ladybird, pupa

***Ladybirds on bark***

DAVID ELEMENT

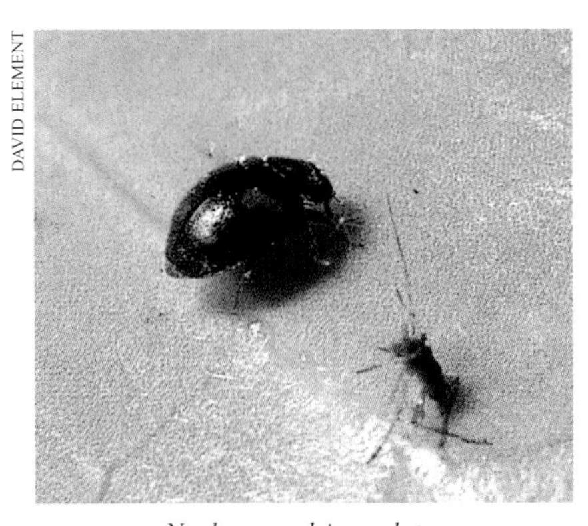

*Nephus quadrimaculatus*

GRAHAM COLLINS

*Rhyzobius litura*

GRAHAM COLLINS

*Hyperaspis pseudopustulata*

GRAHAM COLLINS

*Rhyzobius litura* – larva

GRAHAM COLLINS

*Scymnus frontalis*

ROBERT POPE

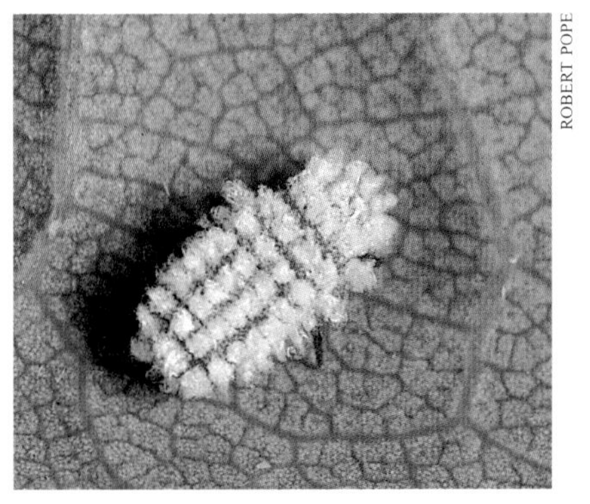

*Scymnus auritus* – larva

***Smaller ladybirds***

GRAHAM COLLINS

*Anatis ocellata* – Eyed Ladybird

GRAHAM COLLINS

*Harmonia 4-punctata* – Cream-streaked Ladybird

GRAHAM COLLINS

*Myzia oblongoguttata* – Striped Ladybird

GRAHAM COLLINS

*Aphidecta obliterata* – Larch Ladybird

GRAHAM COLLINS

*Exochomus quadripustulatus* – Pine Ladybird

GRAHAM COLLINS

*Chilocorus renipustulatus* – Kidney-spot Ladybird

***Ladybird larvae***

GRAHAM COLLINS

*Propylea 14-punctata* – 14-spot

DAVID ELEMENT

*Calvia 14-guttata* – Cream-spot Ladybird

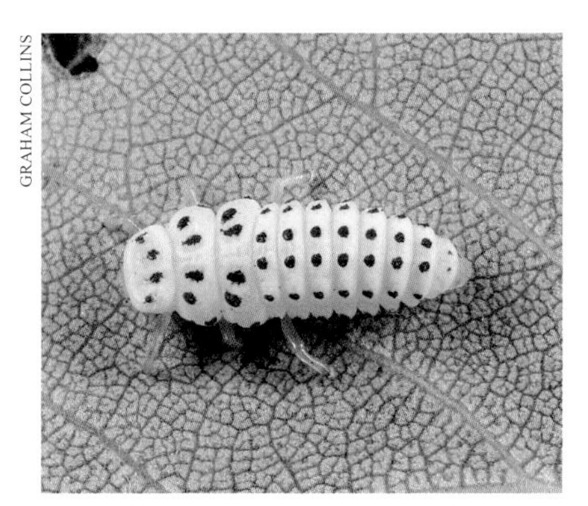

GRAHAM COLLINS

*Halyzia 16-guttata* – Orange Ladybird

MICHAEL MAJERUS

*Psyllobora 22-punctata* – 22-spot

GRAHAM COLLINS

*Coccinella 5-punctata* – 5-spot

MICHAEL MAJERUS

*Tytthaspis 16-punctata* – 16-spot

***Ladybird larvae***

GRAHAM COLLINS

*Coccinella 5-punctata* – 5-spot

GRAHAM COLLINS

*Coccinella 11-punctata* – 11-spot

DAVID ELEMENT

*Chrysomela populi* – a leaf beetle

DAVID ELEMENT

*Lilioceris lilii* – Lily Beetle

DAVID ELEMENT

*Phytodecta viminalis* – a leaf beetle

DAVID ELEMENT

*Apoderus coryli* – a leaf-rolling weevil

***Two rare ladybirds, and other red beetles***

pairs on each side. These are on the two outermost tubercles of the first and fourth abdominal segments. The head and first thoracic segment also have some orange marks.

The year of the great ladybird plague, 1976, is now fast fading into history. There was virtually no rain in Surrey for three months of this long, hot summer. The grass turned brown and insect life disappeared, except for enormous numbers of 7-spot Ladybirds. They migrated to find better conditions and became concentrated on the coast and a nuisance to holidaymakers, even biting them in their desperation for food and water. Many aspects of this remarkable event are described in Majerus (1994). In Surrey, although 7-spots were abundant, I did not experience them biting.

There has been much speculation about the cause of the huge numbers of 1976, but this is not entirely clear. The most plausible explanation is given by Majerus (1994). He states that a successful breeding season in 1975 was followed by a mild winter, with abnormal numbers of adult ladybirds surviving it. In spring these met an abundance of aphids, which had also survived the winter well and were multiplying rapidly in the warm, wet weather of early 1976. The large numbers of aphids and the hot dry weather of June led to another highly favourable breeding cycle producing a new generation of ladybirds in massive numbers around midsummer. As the dry weather persisted and plants withered, aphid numbers crashed and billions of hungry ladybirds began to roam the countryside and arouse so much interest in their unusual behaviour. Other writers have put forward the alternative theory that the 7-spot Ladybird produced two generations in the spring of 1976 but, without evidence to support it, this has to remain speculation.

One reason for keeping a detailed diary of ladybird numbers was to plot the development of the next ladybird plague but, after 20 years of note-taking, this has not yet occurred. It has been anticipated once or twice, particularly in 1990 when numbers were high in a mild, early spring, but these conditions did not lead to a summer population much greater than normal. Although mild winters have been frequent and hot summers have become the rule rather than the exception, the conditions of 1976 have not been repeated.

Instead of a population explosion, there have been more moderate fluctuations and finally a crash in numbers, so that the survey closed with the population of 7-spot Ladybirds at an abnormally low level. This is not a new phenomenon, since only 15 specimens (0.5% of total ladybirds) were seen around Kew between 1963 and 1968, following the bitter winter of 1963 (Eastop & Pope, 1969). More recently, on 7 April 1995, a walk through this area from Putney to Richmond produced 41 7-spots in a single day (12% of total ladybirds), although the species was still outnumbered by the 2-spot (16%) and Pine Ladybird (66%).

Numbers at the start of the survey seemed quite high, with records of 20 to 60 at a single locality being commonplace in 1984. The species was somewhat less abundant in the years that followed, with noticeably few in 1988 compared with other ladybirds, but there were high numbers again in 1989 and particularly in 1990. From then on, recording was irregular but the 7-spot remained a common species until 1999 when, following a mild early spring, the first three weeks of April were cool and wet while the last week was fine but cold with night frosts. On 2 May just two specimens were found in a full day's recording and I concluded that most of the overwintered generation had perished. Some eggs must have been laid, since a few fresh adults came through in late July. During the autumn the species was widespread but still in very low numbers and it was now a regular occurrence to walk for a day and see just two ladybirds, or even none at all on 18 October. In a normal year it would be unthinkable and probably impossible to keep count of all the 7-spot Ladybirds one met, but I did just this in 1999. In 70 days out-of-doors I counted 180 ladybirds, and three of these were parasitised.

It is a distressing fact of insect life that many insects have other insects living inside them. Ladybirds are no exceptions and have several internal parasites. One in particular brings itself to our notice by emerging from the body of the ladybird and pupating within a fluffy brown cocoon beneath it. The ladybird is still alive but partly paralysed so it cannot walk away, and the parasite gains some protection from its presence. This parasite, a small black wasp named *Dinocampus coccinellae* (= *Perilitus coccinellae*), has been reported as affecting 17% of 7-spots in East Anglia in 1986/7 (Majerus, 1997) and 59% of samples from eastern Scotland in 1996 (Geoghehan, Thomas & Majerus, 1997). Its numbers in Surrey seem to be low but field observations of parasite cocoons represent the tip of a somewhat unsavoury iceberg, since other ladybirds may still have parasitic larvae inside them. The wasp may be increasing in Surrey, since our records date from 1990, 1997 (3), 1998 and 1999 (3). The records are from April, May (4), August, September and October, suggesting that the wasp has two or even three generations in a year.

The 7-spot almost invariably has just a single generation a year in Surrey. Mating pairs are usually seen in April and May, but mating activity began in March in the early spring of 1990 and continued into June in the late spring of 1984. Larvae and pupae can often be found in large numbers, particularly in June. The first fresh adults, distinguished by their orange background colour, appear in mid-June in an average year but not until July in some years. Some precocious individuals emerged in late May in 1992 and 1995. The new adults feed up for a while but may become dormant at the height of the late summer warmth – there were many small clusters on gorse and pine at Blackheath near Guildford on 19 August 1984. In this species it seems that aestivation may run on into hibernation. The possibility of a minuscule second generation cannot be ruled out. A mating pair on 20 September 1986 was probably too late in the year for its offspring to be successful, but two larvae on 14 September 1986 and another on 2 September 1998 may well belong to a second generation.

In winter 7-spots may hide among the foliage of scale-leaved conifers or sit exposed on the shoots of gorse bushes, sheltering between the spines, or on the shoots of young pine trees between the needles, but probably a greater part of the population hibernates down in the leaf litter, sheltering typically in curled dead leaves. In spring and summer the ladybirds feed on a great variety of aphids on many different kinds of flowering plants, both low-growing ones such as yarrow, ragwort, red dead-nettle and meadow vetchling, and tall herbaceous plants such as nettles, mugwort, cow parsley, hogweed and other umbellifers, where they sometimes also feed at the nectaries of the flowers. On trees the 7-spot generally keeps close to the ground and often sits near the base of a trunk even when other ladybirds such as 2-spots are roaming about higher up, but it occasionally breeds on bushes such as gorse, broom, sallow and even small pine trees.

Some years ago I was fascinated to hear an account of large numbers of ladybirds on a hilltop in North Wales; a similar occurrence near the summit of Snowdon has been reported by Majerus (1994). Ladybirds of several species, including 7-spots, frequently pass the winter in large clusters on stony mountain-tops in parts of Central Europe (Hodek, 1973), but this habit has rarely been reported from this country. Surrey is a lowland county and our highest point is Leith Hill at 295 metres. Just below the top of this hill, at Coldharbour, there were 131 7-spot Ladybirds on various low shrubs on 14 April 1984, clearly not long out of hibernation. This is perhaps not the most conclusive evidence, but the idea that ladybirds may occasionally congregate in winter on the tops of Surrey hills, relatively low as they are, remains an intriguing possibility.

## *Coccinella magnifica* Redtenbacher, 1843 PLATE 4 **Scarce 7-spot Ladybird**

(= *Coccinella distincta* Faldermann, 1837)

**National Status:** Notable (NA)
**Number of Tetrads:** 21
**Status in Surrey:** Very local
**Habitat:** Near nests of wood ants

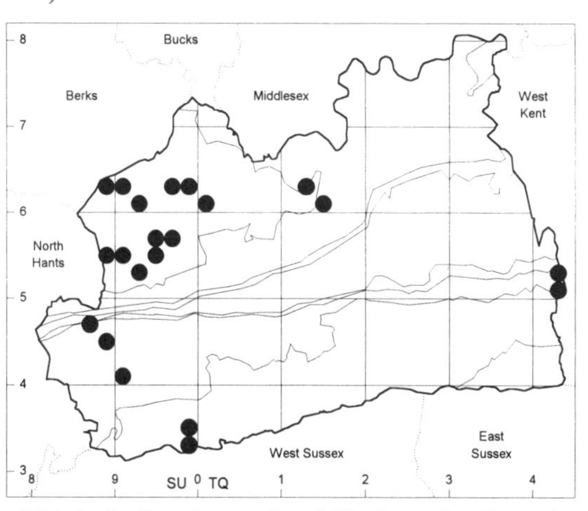


The Scarce 7-spot has also been called the Ants'-nest Ladybird (e.g. Burton, 1968) and the two names together give the essential facts about this interesting species. It is closely similar to the common 7-spot Ladybird but is a very much rarer species because it is only found within a few metres of the nests of the wood ant, *Formica rufa*. This is the ferocious red-and-black ant that forms its nests as great mounds of dead sticks and other debris. Most colonies of wood ants are in light woodland on sandy soil and so are concentrated in the west of the county. It has been difficult for our recorders from east Surrey to do more than make occasional visits to these areas and document the presence of the ladybird. Fortunately, this ladybird has attracted the attention of researchers from Cambridge University so we now know more about the reasons for its strange behaviour (Majerus, 1989; Sloggett *et al.*, 1999).

The Scarce 7-spot differs from the common species in several different ways, even though all these differences are slight. The most obvious is in the relative sizes of the spots. The common 7-spot has all its seven spots more-or-less equal in size, but the rare species has the central spot on each wing-case comparatively large and the foremost spot comparatively small. On occasions I have used this character to pick out the Scarce 7-spot from among ordinary 7-spots without realising that wood ants were present, but one can also be misled by a common 7-spot with unusually dissimilar spots.

In cases like this, other features need to be checked and two in particular are useful. When viewed in profile, the wing-case of the common 7-spot slopes gradually towards its tip, while that of the Scarce 7-spot falls away abruptly, giving it a square-ended look. In motoring terms, the common 7-spot is an estate car while the Scarce 7-spot is a hatchback. On the underside of the body, the common species always has just a single pair of white spots outside the base of the middle legs, while the Scarce 7-spot has an extra pair of white spots in line with the base of the hind legs (*see Figure 2 on page 13*). These additional spots are often small and faint compared with the front pair, so this character needs to be used with caution. Even if faint, however, the extra spots are always present in the Scarce 7-spot.

The above characters are always sufficient to separate the two species, but other differences exist that can be seen when specimens are placed side-by-side. When carefully examined, the side edges of the wing-cases splay out slightly in their front half on the common 7-spot but not on the other species. The front angles of the pronotum project a little in the common 7-spot but are more rounded in the Scarce 7-spot. On the underside of the pronotum, the white colour at the front angle is more extensive in the Scarce 7-spot and occupies at least half its length.

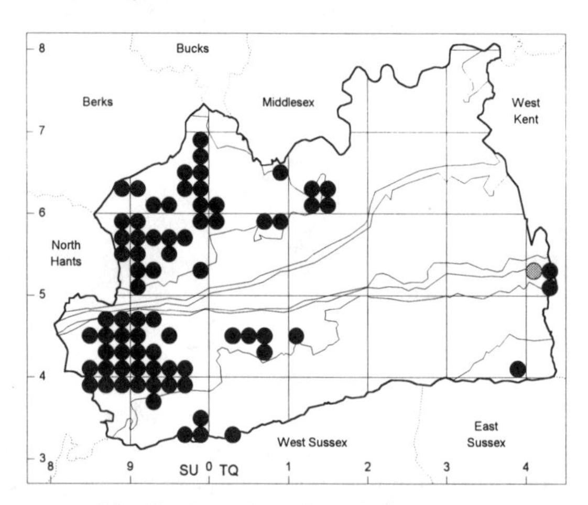


***Distribution of wood ant,* Formica rufa**

A comparison of the distribution map for the Scarce 7-spot with that for the wood ant, *Formica rufa*, shows that the ant is more widespread in Surrey. The ladybird is usually present when the wood ant is well-established and abundant, but we have searched hard and failed to find it in some localities with very few ants' nests, either where the nests appeared to be of recent origin, or where the ants were declining because of destruction of their habitat or changes to it. Since the ladybird is not as easy to record as the ant, it may well yet be found at some sites additional to those shown on the map.

We have found the ladybird in bushes and on trees near ants' nests, sometimes directly above the nest but very rarely actually on it. Sometimes common 7-spots also occur close to the ants' nests, and then the two species require careful separation. Observations at one site give some idea of the life-history of the Scarce 7-spot. On 4 June 1988 I found about a dozen ladybirds crawling on the ground and on low vegetation, together with a larva.

Taken into captivity, this larva pupated on 10 June and the adult emerged on 23 June. Returning to the site on 28 August, there were many adults of the new generation (with orange background colour) on young trees near the ants' nests. At another site on 24 September 1989, many 7-spot Ladybirds were grouped into small clusters at the ends of branches of young pines. Two of these clusters were Scarce 7-spots, with about six ladybirds in each. It is interesting that the two species did not mix on this occasion.

The life-history of the Scarce 7-spot and its association with wood ants was described by Donisthorpe (1919-20) and discussed and expanded by Majerus (1989). The ladybird never goes into the ants' nest but spends its entire life a few metres away, feeding on the aphids tended by the ants. Other ladybirds are attacked by the ants and so are rare in the vicinity of their nests, but this species is tolerated, perhaps because it releases a chemical that pacifies the ants.

Our only record of Scarce 7-spots not immediately associated with *Formica rufa* was made by John Pontin at Sheets Heath on 15 June 1990. Here the ladybird was among aphids tended by a different ant, *Formica sanguinea*, the so-called slave-making ant which does not form nest mounds like *F. rufa* but makes small nests, often under dead logs. This same association had been observed previously in Worcestershire (Pontin, 1960).

The Scarce 7-spot can be seen at the Surrey Wildlife Trust's nature reserve of Brentmoor Heath. The wood ant, together with the ladybird, seems to thrive at sites such as Bagshot Heath and Esher Common where trees are cut back regularly along forest rides or beneath electrical power lines. This suggests that a coppicing regime might be the most appropriate management for the conservation of wood ants, the Scarce 7-spot Ladybird and all the other specialised insects that live alongside the ants.

## *Coccinella hieroglyphica* Linnaeus, 1758 PLATE 11 **Hieroglyphic Ladybird**

**National Status:** Local
**Number of Tetrads:** 26
**Status in Surrey:** Very local
**Habitat:** Heathland

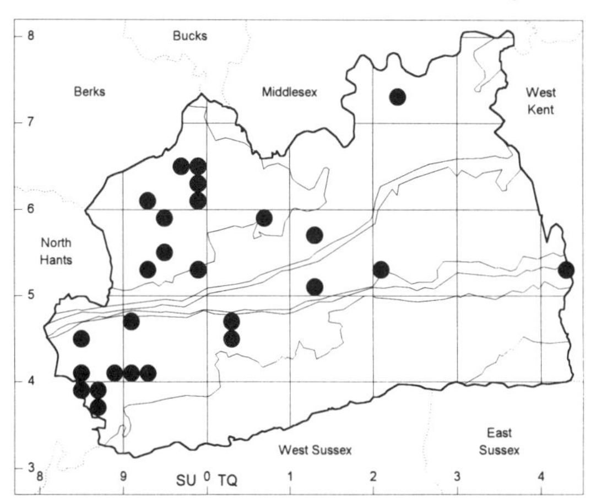


In this species the spots on the wing-cases are joined wholly or partly by longitudinal lines, giving a fancied resemblance to an ancient Egyptian hieroglyphic character – was the author, Linnaeus, thinking of a particular hieroglyph or just using his imagination? In the typical form with few black spots on an orange-red background, the single spot at the base of the wing-cases is always elongate. The frequent melanic form has the whole of the wing-cases black except for a few obscure pale marks, usually around the edges. A chequered or intermediate form is less common. In this survey the variety has rarely been noted, but Majerus (1994) gives the proportions as two-thirds typical, one-quarter melanic and less than one-tenth intermediate. This ratio applies equally to Chobham Common in Surrey and to other sites in southern England.

In all forms of the Hieroglyphic Ladybird, the underside and legs are black while the pronotum is black with white spots at its front angles. The melanic and intermediate forms never have the broad red shoulder-spot that is present in the corresponding forms of the 2-spot and 10-spot Ladybirds.

The larva resembles a small version of the 7-spot but it is black and the pairs of pale spots on each side of abdominal segments 1 and 4 are whitish rather than orange. It has not been found during this survey. Its food was long considered to consist solely of the heather aphid, *Aphis callunae*, but recent research and observations have shown that it can also feed on the eggs and larvae of certain leaf-beetles (Chrysomelidae), including the heather beetle, *Lochmaea suturalis*.

The habitat and appearance of the heather aphid was described by Stroyan (1984). It is a small, dark brown aphid, often appearing pinkish through a covering of powdery wax. It lives on old straggling plants of heather that are typical of woodland clearings and margins, degenerate heathland invaded by scrub, or roadside verges. It is a cryptic species, closely resembling the dead florets of the heather, and shams dead when beaten from the plant. Searching for this aphid may be a useful first step towards finding the larva of the ladybird and studying its feeding preferences.

This habitat description applies to many of the sites where we have found the Hieroglyphic Ladybird. Most of these are old heathland being invaded by pine and other trees, or even, in the case of Limpsfield Chart, planted with conifers. In pure stands of heather we have often searched in vain. At Chobham Common in 1988, it took 20 minutes of sweeping to find a single specimen in apparently prime habitat (VW). At sites where heathland is being created or restored, such as the Surrey Wildlife Trust's reserves at Nower Wood and Bagmoor Common, the ladybird is absent although the heather beetle is common.

Although the Hieroglyphic Ladybird appears to be declining along with the heathland, it is still present at some outlying sites: Bookham and Wimbledon Commons (MVLB), Headley Heath and Ranmore Common (KNAA), and Limpsfield Chart where both the heather and the ladybird have so far survived in the rides alongside the plantations, and made a comeback when the trees were felled.

## [*Coccinella 5-punctata* Linnaeus, 1758 PLATES 15, 16 **5-spot Ladybird**

The 5-spot Ladybird has never occurred in Surrey but photographs of the adult and larva are included in this book as a help to those people who might be contemplating a survey of their own in another part of Britain. In recent years this species has only been recorded from banks of shingle by rivers in west Wales, Scotland and possibly Devon, but it formerly occurred in other counties of western and northern England, and in Europe is widespread in more varied habitats (Majerus, 1994).

The adult resembles a small version of the 7-spot Ladybird with the front spot on each wing-case absent and the back spot reduced in size. The larva is immediately recognisable from the additional orange spots on its abdomen. In addition to the pair of spots on each side of segments 1 and 4, as in the 7-spot, the middle tubercle on each side of segments 6 and 7 is also orange.]

## *Coccinella 11-punctata* Linnaeus, 1758 PLATE 16 **11-spot Ladybird**

**National Status:** Local
**Number of Tetrads:** 31
**Status in Surrey:** Very local, probably migrant
**Habitat:** Low vegetation, usually on sandy soil

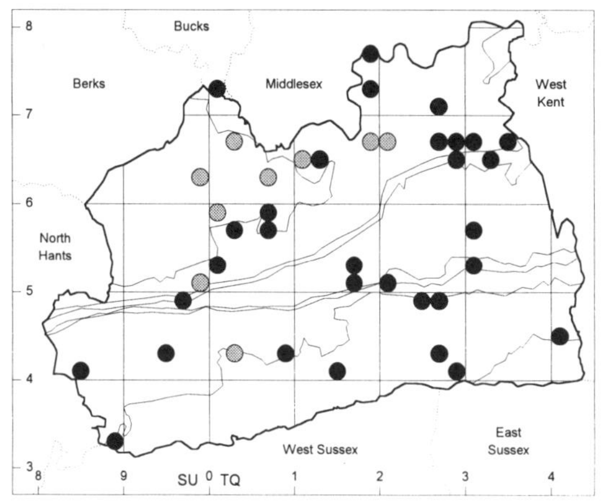


Among the hordes of 7-spot Ladybirds there may occasionally be found a smaller, narrower species about 4.5 mm long. This often turns out to be an 11-spot, but it needs to be separated carefully from the somewhat similar Adonis Ladybird. The pattern is fairly constant with eleven black spots on a red background, with a touch of white beside the basal spot. As in related species, this background colour is orange-red in young adults. The five spots on each wing-case are arranged in a central row of two and an outer row of three, but one of our records refers to a nine-spotted form with the front spot missing from the outer row.

The Adonis Ladybird may also have nine or eleven spots disposed in a similar manner, but the pronotum is different. On the 11-spot it is broadest at the base (broadest in the middle on the Adonis) and black with broad white front angles in contrast to the four-fingered or spectacle pattern of the Adonis. The underside and legs are completely black apart from two white spots outside the base of the middle legs.

This is an elusive species that can turn up anywhere but seems to have no permanent colony in Surrey. As befits a species that is more common in coastal districts, it is most often found in sandy areas such as Mitcham Common, Wisley and south of Reigate. At a water-filled former sandpit near Merstham in 1998, the presence of this ladybird combined with a discarded beach-towel and a bracing wind whipping up the waves on the lake to create a distinct impression of being at the seaside.

The larva resembles a small example of the 7-spot with orange spots in pairs on a grey-black background, but the early stages have not been reported during this survey. Occasional breeding is suggested from finding two together, both with orange background, at Gatwick Airport on 11 July 1984. All other records were of ones and twos, except for three tapped from dead thistles along Halfway Lane just outside Godalming on 21 October 1990. Six

Adonis Ladybirds were also present on the thistles along the sandy banks of this farm track but, on returning to the site in May 1998, neither of these species could be found.

The 11-spot Ladybird often comes into houses in areas where it is common, but with so few outdoor records I was surprised to find one in my home in Horley on 7 February 1995. Other records of overwintering indoors come from a house in Addiscombe on 8 March 1986 and even on the exotic shrub *Protea* in the Temperate House of Kew Gardens on 6 January 1990 (both JMcL). It can apparently overwinter successfully out-of-doors, for one was among the foliage of Lawson's cypress in Guildford on 9 March 1985.

A thorough investigation into the life-history and distribution of this species was made in 1969 by Benham and Muggleton (1970). They confirmed that it was an aphid-feeder, contrary to one previous suggestion that the larvae developed in cow dung. Its distribution was predominantly coastal in northern and western Britain but it also occurred widely over a large part of southern and midland England, particularly in areas of sandy soil. A link with regions of low humidity was suggested. They found viable populations at sites in Middlesex, but the frequency in this neighbouring county has now reduced to be comparable with that in Surrey, whereby a recorder might see just one or two specimens in a year (John Muggleton, *pers. comm.*). Since this comparative rarity in Surrey has been evident throughout the 1980's and 90's, a population decrease in the 1970's is implied. One major climatic event of this period was the drought of 1975/76 in which the 7-spot Ladybird took centre stage, but numbers of 11-spots were also unusually high at this time (Majerus, 1994).

Subsequent recording has shown that the 11-spot also occurs inland in other regions, including the Scottish Lowlands (Majerus *et al.*, 1997). One difficulty of relying on 10-km distribution maps is shown by the effect of placing our records on such a map. It would appear to be a common species from being recorded in 17 10-km squares in Surrey, but only 41 specimens have been seen in 20 years of rather intensive recording, which puts it among our rarest species. My personal experience of the 11-spot outside Surrey has been of finding it to be common near the coast of Kent, by the Severn Estuary and on sandy coasts in Pembrokeshire, which I contrast with its rarity in Surrey. A detailed and systematic survey of a county with a coastline would be very interesting.

The reason for the apparent preference of the 11-spot for coastal areas is open to speculation and I would like to see its national distribution surveyed in the year following a severe winter. This was done in Surrey following the bad winter of 1962/3, when two specimens were found around Kew during the period 1963-5 and one more during 1966-8 (Eastop & Pope, 1966, 1969). These numbers are comparable with the present survey. Our records are unevenly distributed over the years, as shown in the following table which also indicates the severity of the preceding winter.

| 1980 | '81 | '82 | '83 | '84 | '85 | '86 | '87 | '88 | '89 | '90 | '91 | '92 | '93 | '94 | '95 | '96 | '97 | '98 | '99 |
|---|---|---|---|---|---|---|---|---|---|---|---|---|---|---|---|---|---|---|---|
| 1 | 1 | ** | 1 | 8 | 1 * | 1 ** | * | | | 9 | 2 * | 1 | 1 | 3 | 5 | 1 * | * | 3 | 3 |

* at least one week of frost in early part of year    ** over three weeks of frost

***11-spot Ladybird – number of specimens recorded each year***

## *Harmonia quadripunctata* (Pontoppidan, 1763) PLATES 10, 14

## Cream-streaked Ladybird

**National Status:** Local
**Number of Tetrads:** 97
**Status in Surrey:** Local
**Habitat:** Restricted to Scots pine and black pine

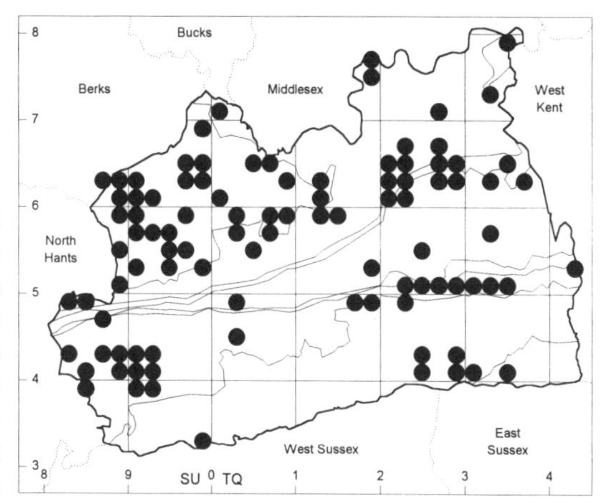


The Cream-streaked Ladybird is the most widespread of the three large species breeding on pine trees, although it has only been known in Britain since 1939. The cream streaks of its name refer to the background colour of the wing-cases, which varies from a greenish-yellow in fresh adults to an uneven orange-red with pale streaks in older ones. The scientific name refers to a common variety with four black spots on the outer sides of the wing-cases, but a still more common strongly-spotted form has at least sixteen black spots arranged in a 1-3-3-1 pattern on each wing-case. The varieties are not always easy to separate since the additional spots are sometimes faint, but we recorded the variety for 32 specimens of which 75% were strongly sixteen-spotted and a further 9% more faintly so.

The four-spotted variant of the Cream-streaked resembles a form of the 10-spot Ladybird, which is a smaller insect with less white on the underside and a different pattern on the pronotum. That of the 10-spot has a semicircle of four black spots surrounding a central spot, while the pronotum of the Cream-streaked has two inverted triangles of small black spots outside this semicircle. On its underside there are four white spots linked by a white bar, compared with only two white spots beneath the middle legs in the 10-spot Ladybird.

The larva is a most distinctive creature with an orange line on its side that is clearly visible even to the naked eye at some distance. This line is formed by bright orange spots on the middle tubercles of the first four abdominal segments. In addition to this, the larva has a unique structural feature. The spines on each tubercle are branched into three prongs which splay outwards, giving it the appearance of a miniature Stegosaurus.

The Cream-streaked Ladybird was first found in Britain in Suffolk in September 1939 on larch, and two years later on spruce. It was found on pine in 1943 in Berkshire (modern Oxfordshire) and again in 1948 in Cambridgeshire. The first Surrey record was of a single example beaten from pine at Oxshott Heath on 9 September 1956 (Buck, 1957). It was already quite a common species when I first looked at pines in 1984 and has since become increasingly widespread. It now ranks second in abundance among ladybirds on pine.

It is found on Scots pine in many situations, not only in the extensive pine woods on the west Surrey commons, but also on isolated trees or small groups in parks, cemeteries, gardens and roadside plantings. Almost all our records of larvae, pupae and freshly-moulted adults came from Scots pine, but four records from black pine of up to ten adults, including a mating pair on 17 March 1990 and a fresh adult on an isolated tree at Oxshott in 1997, indicate that it is also breeding on this tree.

It is often beaten from other trees immediately adjacent to pines, but only rarely found away from its natural habitat. Several were on other conifers such as Norway spruce and Douglas fir, mostly in May and September but with no evidence of breeding. Two found among the sprays of scale-leaved conifers in gardens, including one on Sawara cypress on 1 February 1985, suggest overwintering among their foliage rather than breeding.

Our only record of feeding is of a larva eating ladybird eggs, not necessarily of its own species, but a fresh adult was found on Scots pine with some large brown aphids which it happily ate in captivity.

Larvae and pupae occur commonly in the second half of July and through most of August. Our earliest records of fresh adults are 25 June 1989 and 29 June 1997, both early seasons. A late larva was brought into captivity on 31 August 1998, producing an adult on 10 September. Since adults, eggs, larvae and pupae can be found together in July, a partial second generation cannot be ruled out, but we lack observations on mating apart from the pair in March mentioned above.

## *Myrrha 18-guttata* (Linnaeus, 1758) PLATE 10 **18-spot Ladybird**

**National Status:** Local
**Number of Tetrads:** 69
**Status in Surrey:** Local
**Habitat:** Restricted to Scots pine and possibly black pine

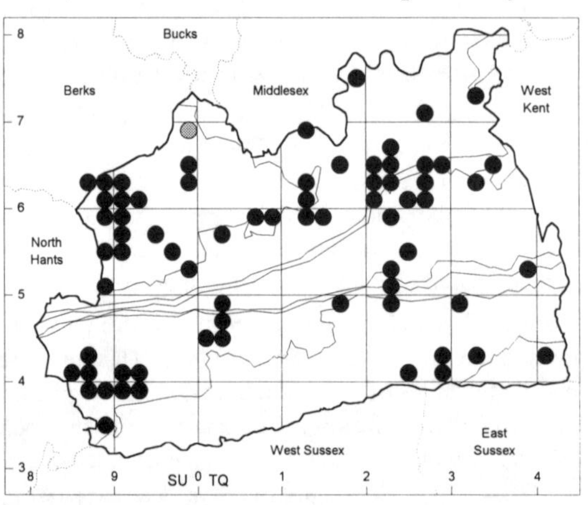


The 18-spot is another ladybird living only on pine trees. It is rather small (4 mm long) and a rich deep brown with the eighteen spots having a silvery tinge. It is most similar to the Cream-spot Ladybird and at least two of our recorders have reported confusing these two species. They are best separated by the shape of the spot closest to the base of the wing-cases. On the 18-spot this spot is L-shaped with the base of the 'L' facing inwards and forwards. The Cream-spot has all its fourteen spots more-or-less round or oval.

The dark mark on the pronotum is in the shape of a curly 'M' with broad outer legs and a narrow central one. It may be likened to a creature looking over a wall – a creature with big ears and narrow, close-set eyes – what it is depends on the pattern of the individual specimen and the reader's imagination. These small 'eyes' are absent in the Cream-spot which may instead have a pale central mark. The underside of the 18-spot is a slightly darker brown than the upperside. Running back from the white spot beside the base of the middle legs is a white bar along the side of the thorax (absent in the Cream-spot).

This species has long been considered to live on the higher parts of pine trees (no doubt following its prey) and this has been demonstrated by counting specimens on recently-fallen pines in October (Majerus, 1988). However, sufficient examples must have gravitated towards the lowest branches to give us a very reasonable map of its distribution. Of course it needs only a single specimen to give a dot on the map.

The larvae and other early stages were found by Michael Majerus in June and July 1989, principally in the top six metres of the crowns of mature pine trees (Majerus, 1994, p.154). Some of these observations were made at Chobham Common in Surrey, but the larvae have otherwise not been recorded at all in this survey. If found, the larva would resemble that of the Larch Ladybird.

In Surrey the 18-spot is the third most widely-spread of the ladybirds on pines, after the Pine and Cream-streaked Ladybirds. It often occurs on isolated trees in parks, gardens and cemeteries as well as on the abundant pines of the western heaths. It can even be found on quite small trees no more than four metres in height. The dates of our records (chart below) show peaks in May and August and fit in well with the direct observations of larvae and pupae mentioned above. Out of 100 individuals whose origin was recorded, 90 came from Scots pine and two from oaks adjoining pines or growing below them. A further eight were found on black pine at four separate localities including a churchyard and two cemeteries, which suggests that the 18-spot is also breeding on this widely planted tree, but observations of larvae or pupae are required to prove it. Almost all our records came from beating the pine trees, but the few ladybirds found by searching were either on the cones, needles or small shoots.

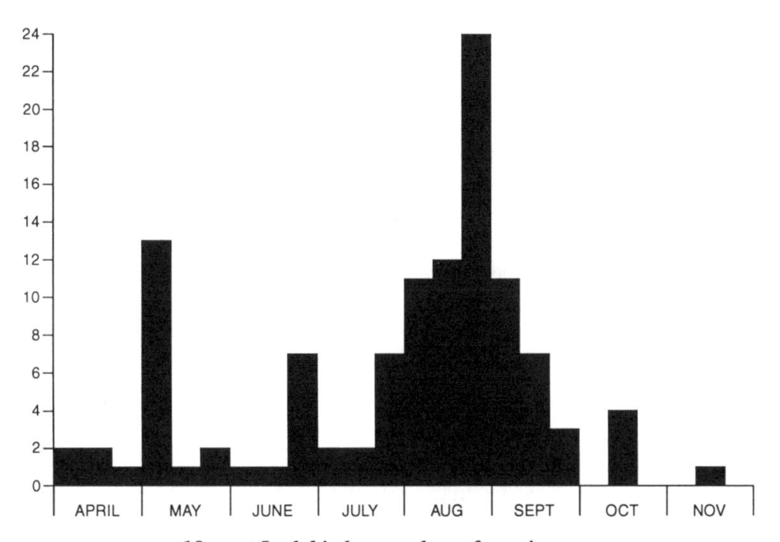


***18-spot Ladybird – number of specimens***

One overwintering example was found in a grass tuft at the foot of a pine tree on 11 March 2000, but others have been beaten from pines in October and November. This species sometimes flies to mercury-vapour light, appearing on three occasions at South Croydon although 70 metres from the nearest pine (GAC).

## *Calvia 14-guttata* (Linnaeus, 1758) PLATES 7, 15 **Cream-spot Ladybird**

**National Status:** Common
**Number of Tetrads:** 168
**Status in Surrey:** Common
**Habitat:** Ash, lime and other broad-leaved trees

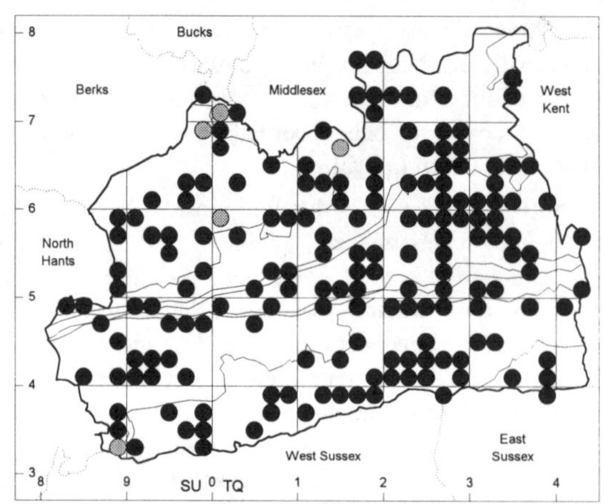


The Cream-spot Ladybird lives normally among the foliage of broad-leaved trees but often descends from its natural habitat and rests on the leaves of wayside plants inviting our attention. It is then one of the first species to be noticed by the amateur naturalist as a ladybird with a difference. Instead of the black spots on red of a normal ladybird, the spots are creamy white on a chestnut-brown background, which can be orange-brown in fresh specimens. There are almost always fourteen spots on the wing-cases as in the scientific name. This is a medium-sized species (about 5 mm long). The pronotum is the same colour as the wing-cases with pale spots at its rear angles. The head, legs and underside of the body also match the upperside in colour, apart from a broad yellow triangle outside the middle legs.

Within the British fauna, confusion is only possible with the 18-spot and Orange Ladybirds. The 18-spot differs in its L-shaped spots at the base of the wing-cases and lives in a different habitat, on coniferous trees. The habitat, however, cannot be relied upon and at least two of our recorders have admitted to an initial difficulty in distinguishing this species from the Cream-spot. The Orange Ladybird is normally paler and orange-yellow in colour with whiter spots, but occasional examples of both species may approach the colours of the other, so a quick check of structural detail is necessary. The wing-cases of the Cream-spot lack the splayed-out flanges of the Orange Ladybird, but I prefer to look at the front of the pronotum which is indented in the Cream-spot Ladybird (straight in the Orange Ladybird). Although the head fills this indentation, it cannot be withdrawn beneath the pronotum.

The larva is rather a handsome creature and is dark, almost black, with bold white marks. As such, it can only be confused with the larva of the 14-spot, and both these species have a tiny projection at the rear end of the abdomen. The white outer tubercles of abdominal segments 4-6 of the Cream-spot are tall and pointed (rounded on the 14-spot), giving a saw-toothed appearance to the side of the body, and it also lacks the broad white central spot on the fourth abdominal segment of the other species. The white stripe down the centre of the thorax is often slimmer. When found among larvae of the Pine Ladybird, the Cream-spot impresses as a larger, fast-moving larva with unbranched spines.

The species is common in Surrey and found in all parts of the county. The incomplete distribution map arises from the difficulty of finding suitable trees with foliage within reach. Persistent searching should eventually reveal its presence in most places except for the most built-up parts of inner London.

A firm association with ash trees is one of the surprises to come out of this survey. Some black-spotted, cream-coloured pupae were found in bark crevices on several trees along the Balcombe Road in Horley in 1984. They were examined daily and eventually adult ladybirds emerged, all Cream-spots. This observation was repeated in succeeding years. With the exception of one pupa on oak and one on lime, none were found on the many neighbouring trees of oak, horse-chestnut, hornbeam, lime and sycamore. The same behaviour was noted along Brighton Road in Sutton in 1988, but here the pupae were all in cracks in the bark on trunks of lime trees, with none on the neighbouring horse-chestnuts. The dates of emerging adults and their number were as follows: 9-27 July 1984 (12), 19-26 July 1985 (7), 7-22 July 1986 (12), 16-24 July 1988 (8), 22-23 June 1989 (5) and 4 July 1990 (1). One that pupated on 9 July 1986 emerged on 16 July, but other individual pupae lasted for at least nine days and emerged on the second hot day after a cool spell. The freshly-emerged adults were spotless and cream or pale orange in colour, but all remained at their site of emergence for up to a day, by which time the pale spots had developed. One that stayed an extra day had already developed the brown colour of the mature adult.

The larvae were never found on the trunks earlier in the year and must have developed higher on the trees, among the foliage. Unfortunately, the branches of ash trees tend to rise at their tips and the leaves are nearly always out of reach. We have therefore few other records of larvae on ash but these include one beaten from foliage on 19 June 1988. One was moving rapidly on an ash trunk on 30 June 1984 (and soon pupated in captivity) while another was seen on the leaves eating woolly aphids. Single larvae were also found on various dates on osier and on field maple, among dark aphids and pale yellow psyllids, while a freshly-emerged adult was beaten from sallow.

Mills (1981) reports it as feeding on various species of aphids on lime, apple and birch, but apparently did not examine ash. Investigations in Europe into the prey of this species agree that it takes both aphids and psyllids but with aphids preferred in one area and psyllids in another. The psyllids listed are the common species on apple, elm and alder but none of the ash-feeding aphids or psyllids are mentioned (Iablokoff-Khnzorian, 1982).

The many adults found in spring came principally from hawthorn (23), birch (9), sallow (8), nettles (14) and other herbaceous plants (11), but surprisingly also from various conifers (6). Mating pairs were seen on 2 April 1999, 9 May 1998 and 6 June 1987 (on an ash trunk). Larvae were found between mid-June and mid-July, with the latest freshly-emerged adult on 12 August 1998. After the breeding season, most adults were found on oak (22), ash (8), lime (5), sycamore (4), birch (3) and sallow (3). We have few records of overwintering, but three were beaten from ivy in late September and October, while two were found low down under trees in November and March.

The clear conclusion is that the species has one generation a year and the larvae develop among the foliage of ash, lime and other broad-leaved trees. An unknown but maybe quite high proportion of the larvae descend the trunks to pupate in cracks low down on the tree where the bark is rough. The species appears to feed on both aphids and psyllids, but further investigation is needed as to its preference on its different hosts.

## *Propylea 14-punctata* (Linnaeus, 1758) PLATES 6, 15 **14-spot Ladybird**

**National Status:** Common
**Number of Tetrads:** 357
**Status in Surrey:** Ubiquitous
**Habitat:** Tall herbaceous vegetation

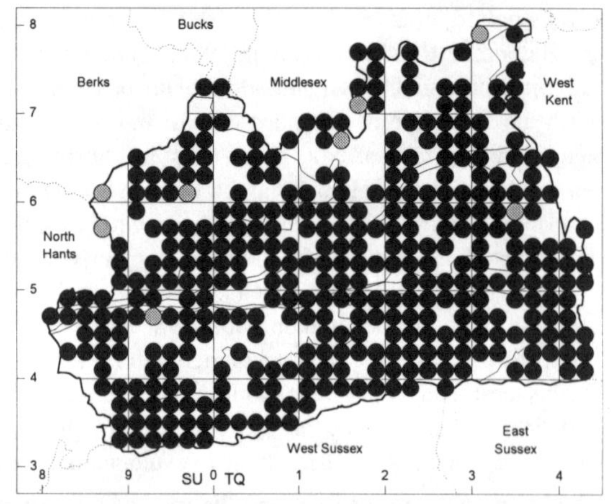


The 14-spot Ladybird is a familiar species with black spots on a creamy-yellow background. The individual spots are rather square and usually link up to form a pattern popularly held to resemble an anchor or even a clown's face. The background colour varies from off-white to distinctly yellow and specimens with separated spots are frequent. This form may be confused with the 22-spot, but its pronotum has two additional spots outside the semicircle of four black spots. The black line along the join of the wing-cases tapers towards the back (distinction from 16-spot). A strongly-marked 14-spot may be confused with the chequered form of the 10-spot, but it has yellow legs with a contrasting black band on the femora of at least the hindmost pair (uniformly brown legs on 10-spot).

The black-and-white larva resembles that of the Cream-spot Ladybird, from which it is distinguished by a transverse white bar linking the topmost tubercles of the fourth abdominal segment. There is often also a white longitudinal stripe that is particularly broad on the thorax, and segments 4-7 of the abdomen have the outer tubercles lower and more rounded than those of the Cream-spot. Both these larvae have a small projecting point at the rear end.

The 14-spot might well be called the dormouse ladybird, since it sleeps for more than half the year! Counts of ladybirds around Kew in west London between 1963 and 1968 showed that more than 99% of specimens were seen in the six months from April to September (Eastop & Pope, 1969). In the current survey, 99.5% of 1,325 seen were found in this same period, and 97% in the five months from 21 April to 20 September. In 1984, the year of most intensive and consistent recording, 14-spots were only seen between 16 May and 2 September, a period of less than four months. The ladybirds must be very well hidden or perhaps very close to the ground for them to elude our searching at other times. There were hardly any 14-spots in the hundreds of ladybirds tapped from grass tussocks in October,

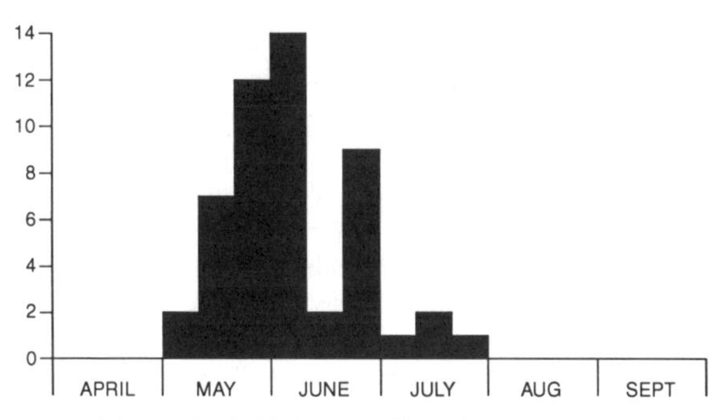


***14-spot Ladybird – number of mating pairs***

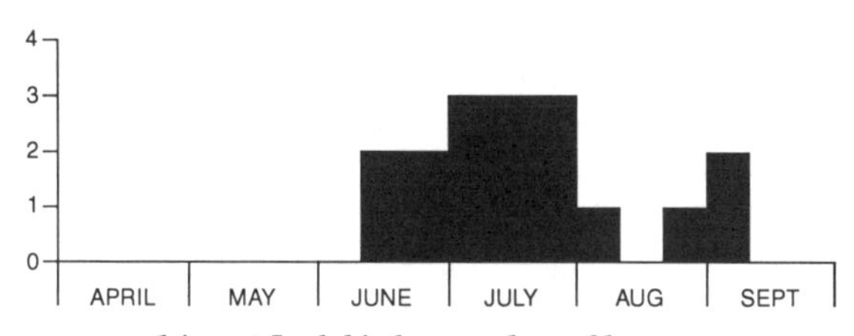


***14-spot Ladybird – number of larvae***

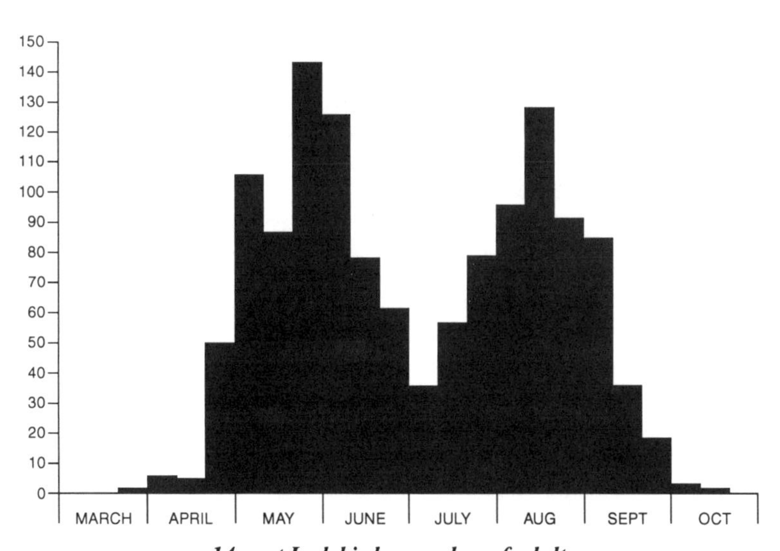


***14-spot Ladybird – number of adults***

but specimens found in moss on 25 March 1996, in grass and leaf litter on 31 March 1990 and beaten from ivy on 23 September 1998 were probably in their overwintering sites.

Even in this short summer period of activity, the species seems able to rear a small second generation. Our charts show a dip in numbers of adults in early July when most of the overwintered generation have died and the new adults have not yet emerged. In 1984 mating pairs were seen in late May and early June (4 pairs), and then another on 21 July at a time when adults of the new generation were emerging. Our colour plate shows a pale example mating with a bright yellow one, but both these are overwintered specimens, since the photograph was taken on 11 May 1994 in Nunhead Cemetery (RAJ). Larvae have been found late in the year on three occasions: 3 September 1988, 1 September 1990 and 23 August 1993. Some were also noted in late June and early July in the first two of these years, and this seems to be their normal season with the September larvae representing a second generation (see chart). Five fully-grown larvae were brought into captivity between 1984 and 1988, including one from September, and all reached the adult stage in 8 to 10 days after spending only 5 to 6 days as pupae.

The 14-spot is widespread in Surrey although rarely found in any great numbers. Our only records of more than ten are of 14 on hogweed and 12 on tansy, both in August 1984. It is likely to prove more common than shown on the map, since some districts may have been surveyed either before or after its limited season. It is however much easier to find in country areas than in heavily built-up parts of London, and is typically found on tall herbaceous plants growing on roadsides or beside hedges.

The ladybirds are usually found on or near colonies of aphids of many different kinds. The habitat was noted for about 1000 individual specimens, only 166 of which came from trees and shrubs, and many of these from low growth, young trees and hedges. Another 51 were on brambles. It was a surprise to find a few specimens on Norway spruce (13) and Scots pine (12), mostly in May. Undoubtedly the most popular host-plant is nettle (191 specimens) which often has its stems and the undersides of its leaves covered with green aphids. Other attractions are colonies of green aphids on hogweed (56 specimens), grey aphids on mugwort (15), and large brown aphids on tansy (18). The ladybird was also reported among aphid colonies on tufted vetch, meadow vetchling, cow parsley, sunflower, teasel, laburnum, broom, rhododendron, sallow, lime, field maple and sycamore. Some were found on bracken (11 specimens), perhaps using its extra-floral nectaries, and there is clearly something of great interest to this species on or around the young fruits of white dead-nettle. Out of 17 specimens found on this plant, seven were probing the calyces surrounding unripe fruits.

## *Myzia oblongoguttata* (Linnaeus, 1758) PLATES 11, 14 **Striped Ladybird**

(= *Neomysia oblongoguttata*)

**National Status:** Local
**Number of Tetrads:** 32
**Status in Surrey:** Very local
**Habitat:** Restricted to Scots pine on the western heaths

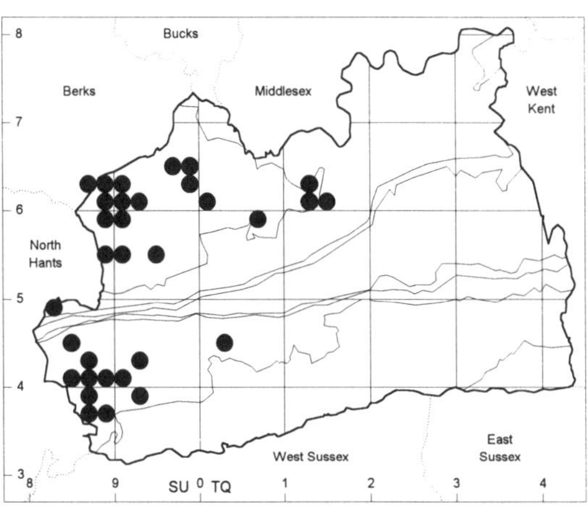


This is a large, handsome species of unusual colour and pattern. The wing-cases are chestnut-brown with pale longitudinal streaks that are somewhat variable but always include a small pale triangle at the base of each wing-case beside the scutellum. The pronotum is a similar colour with broad pale sides and sometimes a pale basal spot. Occasionally the central mark is darker, even black. The underside varies from brown to black with a single pair of white spots outside the middle legs.

The larva is as distinctive as the adult, with a pattern of spots different from that of any other British ladybird. There are orange spots on the two outermost tubercles on each side of segment 1 of the abdomen, and on just the outer tubercles of segments 4 and 6.

The Striped Ladybird is found only on Scots pine and has a more restricted distribution than any of the other large species. It is found only in the pinewoods that have developed on former heathland in north and west Surrey, north-west of a line from Oxshott Heath to

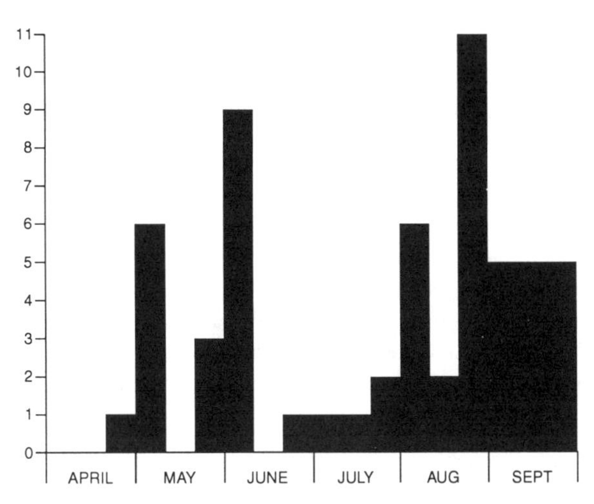


***Striped ladybird – number of adults***

Haslemere. It is also reported from Winterfold Forest, slightly south-east of this line, but we have no precise record.

Even within this limited area it is not very numerous, usually being found only in ones and twos, with only 60 specimens in total being reported to the survey. The most recorded at single sites were six at Lightwater Country Park on 8 August 1998 and seven, including two mating pairs, at Oxshott Heath on 4 June 1988. All specimens were found on Scots pine, and trees with open cones appeared to be favoured, at least in autumn.

The dates of our records suggest a single generation a year, with a first peak of overwintering adults in May and early June, including mating pairs on 28 May and 4 June 1988. Adults are scarce for the remainder of June and most of July, but larvae were found on 19 June 1998 and 30 July 1988. The new generation of adults can be found throughout August and September.

This is a widespread species in suitable habitat throughout Britain and so does not appear to be threatened, but the competition between the large species on pines may be an interesting subject for study, especially now that the Cream-streaked Ladybird, known in the country for just over 60 years, has become more common than the native Eyed and Striped Ladybirds. These three large ladybirds may, however, avoid competition by specialising on different foods.

## *Anatis ocellata* (Linnaeus, 1758) PLATES 11, 14 **Eyed Ladybird**

**National Status:** Common
**Number of Tetrads:** 78
**Status in Surrey:** Local
**Habitat:** Restricted to Scots pine

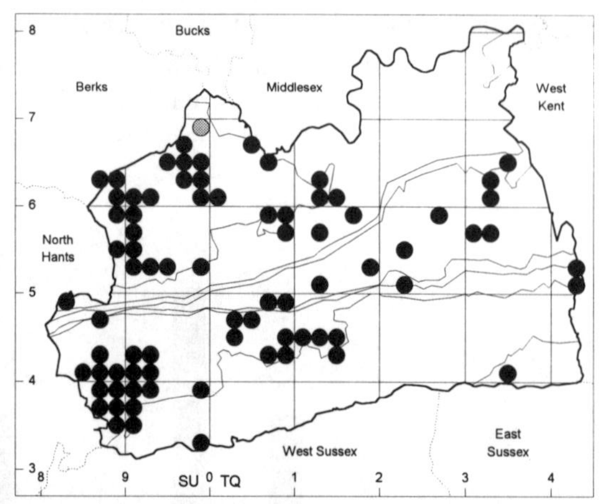


The Eyed Ladybird earns its name from the pale rings around its many black spots. It is our largest species (8 mm) but comparatively little-known because it is largely restricted to pine trees. The wing-cases are red or orange-red and usually have 18 black spots. The pair at the base of the wing-cases are just small elongate wedges that almost touch. The spots vary in size between specimens and each is usually edged by a narrow ring of white or pale yellow. These pale rings were absent from three out of almost 100 adults checked during this survey, but they were very faint on several other specimens.

In very rare cases the black spots may be wholly or partly absent, giving a pale-spotted ladybird, and it is then helpful to recognise the pattern on the pronotum. This has a large black central mark on a white background, with two additional black spots at the sides.

The central mark may be likened to a triumphal arch, top-heavy with two small passageways, or a square-headed masked face peering over a wall.

The larva is also large and its length often exceeds 12 mm. It is easily distinguished from all other large species by having just two prominent orange spots on its side, on the outer tubercles of the first two abdominal segments (Cream-streaked has a row of four orange spots, Striped has additional spots, 7-spot has a pair of spots on segments 1 and 4). The larva also has pairs of small white or yellowish spots along the centre of its abdomen and the remaining outer tubercles may be mostly white. The large pupa is cream with black spots and has two ragged teeth on each side.

It is fortunate that the larva is easily recognisable, since my attempts to rear it have met with a series of disasters – one was thrown out with the dead aphids and old pine needles, while another was left behind the curtains of a bedroom in a German field-study centre.

In Surrey it occurs not only on the pines of the western heaths and in the forestry plantations of the central hills, but on pines in parks and gardens throughout the county. It is not, however, universal in such sites or as common in them as the 18-spot, Cream-streaked and Pine Ladybirds. For instance, I have regularly inspected garden pines in Horley without ever finding an Eyed Ladybird. It is generally found singly or in small numbers, but 25 were counted on Oxshott Heath and Esher Common on 4 June 1988. These included two pairs, the only instance of mating recorded during the survey.

Our observations are wholly compatible with the species having one generation a year. Very few adults were noted during the second half of June and the first half of July, but larvae were recorded from mid-June onwards, with full-grown larvae, pupae and fresh adults numerous towards the end of July and in early August.

Almost 90% of our records of adults came from Scots pine but the breeding habitat is given better by records of the early stages. Ten larvae and four pupae were found on Scots pine but one larva and five pupae were on birch and sallow immediately adjoining pines, although only one adult was found on birch away from pines, and none on sallow. One active larva was moving on grass beneath pine trees. It appears that larvae may often wander away from the tree of their birth, probably when searching for a pupation site.

Two larvae found in Cockshot Wood at Mickleham represent something of a mystery. One came without notes but the other was beaten from a dead larch. The wood was largely broad-leaved and there were spruces nearby but no pines. Most probably the Eyed Ladybird was breeding on larch or spruce at this site.

Most records were made by tapping the branches of pines over an entomologist's beating tray, but a great deal of beating was done to provide few records. Trees with many open cones were most profitable and one adult was found half-hidden between the scales of such a cone on 10 April 1988, but two found on 18 November 1984 were on pine shoots among clusters of Pine Ladybirds.

## *Halyzia 16-guttata* (Linnaeus, 1758) PLATES 7, 15 **Orange Ladybird**

**National Status:** Local

**Number of Tetrads:** 163

**Status in Surrey:** Increasingly common

**Habitat:** Mildewed leaves of certain broad-leaved trees

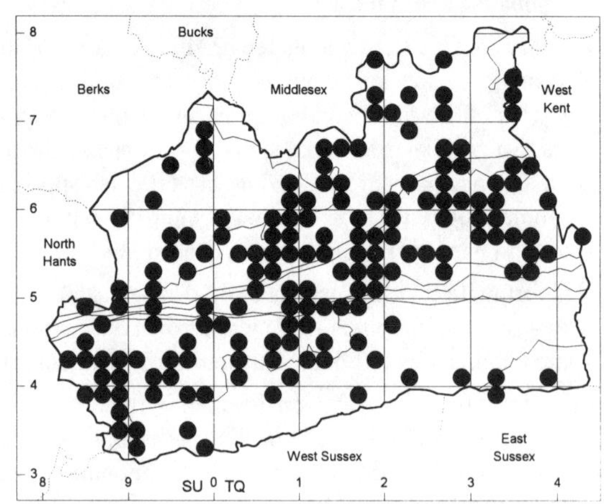


The Orange Ladybird is one of our most attractive species and never fails to draw attention even from the most casual observer. It is distinctive as a larva and pupa as well as an adult, and in its mode of feeding. As its names imply, the adult insect is orange with sixteen white spots on the wing-cases. For many years it was considered to be a rare species of ancient woodland, possibly associated with oak, but during the 1980's and 1990's it has become increasingly common and is now particularly associated with sycamore (an invasive alien in southern England) rather than oak. Because it feeds on white powdery mildew on the leaves of trees, it loves the shady places where this mildew flourishes and so is a true woodland insect rather than one of woodland edges.

The adult can only be confused with young specimens of the Cream-spot Ladybird, but its wing-cases have a distinctive transparent splayed-out edge lacking in the other species. In addition, the front edge of the pronotum is quite straight, without the indentation of the Cream-spot, and the head can be withdrawn below it. Since the pronotum also is partly transparent, the head with its two little black eyes can still be seen.

The larva is delicately coloured. At first sight it appears to be yellow, but closer examination reveals that it is really dull cream with bright yellow longitudinal stripes; there are black dots on all the tubercles. The only other yellow larva is that of the 22-spot Ladybird. Is it a coincidence that both our mildew-feeding species, very different as adults, have similarly coloured larvae? That of the Orange Ladybird has longer legs, its head is a paler brown and the middle tubercle on each side of the first abdominal segment is black-tipped (wholly yellow on the 22-spot). Once the larvae have pupated, the two species are again quite different. The pupa of the Orange Ladybird has a unique pattern, being black with four yellow spots on its back in the shape of a transverse rectangle.

Our understanding of the habits of this species was revolutionised by the discovery in 1987 of its larvae and pupae underneath the leaves of sycamore, feeding on mildew (Majerus & Williams, 1989). One of the first observations of this was made during a course at Juniper Hall Field Centre attended by several Surrey recorders, and this was a most exciting event. Once we knew where to look for it, the number of records grew slowly and steadily and then with a rush during the last two years of the survey when more than half the records depicted on the map were made.

It must not however be thought that this link with sycamore is exclusive. Even during the Juniper Hall courses of 1987 and 1988, larvae and pupae were found on the mildewed leaves of dogwood, with an adult clearly feeding on this mildew. My own first encounter with the species came during an outing of the Horley WEA in Norbury Park on 2 November 1986 when two unfamiliar, distinctively yellow-spotted beetle pupae were found on a fallen ash leaf by one of our party, Les Cook. I took the pupae home, keeping them in the cool part of the refrigerator, and was extremely surprised to find on 7 November that two Orange Ladybirds had emerged while still in the fridge. (This finding was reported by Majerus & Williams as the first record known to them of the early stages of this insect in Britain, but was unfortunately credited to Sussex!) The ash leaf had fallen recently and was still green, but other trees occurred nearby, including beech and sycamore, so one could not rule out the possibility that the larvae had walked onto the fallen leaf to pupate after feeding elsewhere. It was not until 1998 that I could confirm breeding on ash by finding many larvae and pupae on its mildewed leaves at two sites. From 1989 onwards we found the species on heathland far from any sycamores and usually on birch, and in 1993 the early stages were also discovered on its mildewed leaves. In 1998 larvae were also found in numbers on mildewed leaves of both goat willow and field maple.

In total, the confirmed larval feeding records made during this survey are from sycamore (50), birch (8), dogwood (4), ash (3), goat willow (2) and field maple (2). There is a bias in these figures in favour of sycamore, because in many places only sycamore was searched. Single larvae or pupae were also found on other trees and shrubs, namely larch, oak, hazel, hawthorn and cherry laurel. These are not yet considered proven hosts since in all cases the larvae could have wandered off a known host growing nearby. Hawthorn is the most likely to be a regular host since adults have been found on it frequently at Bookham. Oak leaves often become covered with white mildew in the autumn but neither searching nor beating has produced any larvae of this species. Furthermore, larvae offered this mildew in captivity refused to eat it and were only saved from starvation by being transferred to sycamore leaves that were not so obviously mildewed. Another notable absentee from our list is the Norway maple, related to two known hosts and increasingly common in Surrey both as planted and self-sown trees. Its large thin shiny leaves do not however seem to suffer from mildew as much as its two relatives, sycamore and field maple. Majerus & Williams list sycamore, oak, hawthorn, Scots pine, lime and elm as hosts, but these records include overwintering sites and they only record breeding on sycamore.

For much of the survey period this species has probably been most common on the North Downs around Leatherhead and on the sandy hills of the south-west around Haslemere, but it is now widespread in Surrey and can even be found in London parks. It remains uncommon only in farming areas on the low-lying Weald Clay in the south and south-east of the county, where the woodland and hedgerow trees are mostly oaks.

The first records reported to the survey came from the mercury-vapour light-traps used by moth recorders. Orange Ladybirds are found frequently at moth traps and occasionally at street lamps and house windows. The increased numbers of records in the last few years of the survey may in part be due to more frequent searching, but the operation of a garden moth trap is a regular event, so the opinion of moth recorders offers an unbiased confirmation that the species has indeed become more common (DAC, GAC). Other ladybirds also occur at light-traps, but some may just have been disturbed by the light when resting in the vicinity. This is one of the few species that regularly fly at night and come to light. We have also recorded it as active by day on a few occasions, although some of these were in the evening or at dusk.

The Orange Ladybird breeds late in the year, later than any other ladybird with a single generation. Our only two records of mating pairs were both made in July, at Dawcombe nature reserve on 25 July 1992 and Park Downs, Banstead, on 8 July 1998. The earliest larva recorded is a small one on 26 July 1999, also from Park Downs where the species breeds on sycamore and possibly also on dogwood. A fully-grown larva feeding on mildewed cherry laurel at Wisley on 28 July 1998 produced an adult on 7 August, an unusually early date (AS). Most records of larvae date from the second half of August and the first half of September, and we have only two unusual records after the end of this month (detailed below). The first pupae and newly-emerged adults also appear in late August. Adults continued to emerge until 20 October in 1998 but the last emergence noted in 1999 was on 5 October.

At sites where no adult was found, a single larva or pupa was taken into captivity in order to make absolutely certain of its identity. All but two of the 15 larvae and 22 pupae were reared to the adult stage. They were kept out of the sun in an unheated room. Most pupae emerged quickly, the longest taking ten days, while two others collected as larvae came through the pupal stage in about eight days. One larva moulted then survived 23 days in its last larval stage before dying while pupating; it was probably not given enough food.

A larva beaten from larch on 21 October 1990 was reared to adult by mid-November using mildewed sycamore leaves; larch may well be an unsuitable host for larval development. A small larva on Himalayan birch in Battersea Park on 15 October 1999, with adults nearby, poses something of a problem, being so much later than other larvae observed that year. Although no mating pairs were seen in autumn, it may represent an attempt at a second generation and it is tempting to link it with the remarkable record of two live pupae on fallen sycamore leaves in Thetford Forest, Suffolk, on 7 February 1988 (Majerus & Williams, 1989). These were interpreted as overwintering pupae, but they may have come through part of the winter as larvae. Our Surrey records show no inclination to remain for long in the pupal stage. The possibility of an attempted second generation through the winter months is pure speculation at present, but all late observations of the early stages of this species are clearly of the most profound interest.

Our overwintering records are few but varied. One was found on curtains indoors in Horley on 12 December 1996. There were at least five on rhododendron at Shirley in January 1999 (AJW & MJW). One was moving about slowly on the foot of an ash trunk at Mickleham on 30 January 1995. The six March records include specimens beaten from hawthorn and ivy, and found on grass in a churchyard. One beaten from a juniper bush at Merrow Downs on 1 April 1997 was probably still in its overwintering site, as possibly were some beaten from gorse, Scots pine and Douglas fir near Thursley on 5 May 1996. Right at the end of the survey came our first records of clusters of Orange Ladybirds. Nine were found on a statue in Nunhead Cemetery on 1 January 2000 (RAJ) and 21 scattered over the branches of a single birch tree in a Godalming garden on 14 April 2000 (PRW).

In conclusion we have a most attractive insect, quite large and with distinctive larvae and pupae, that can be found in August and September by searching under the mildewed leaves of sycamore and some other trees and shrubs, that flies at night, prefers shady places to sunny ones, and is becoming increasingly common in Surrey. It has been chosen to adorn the cover of this book.

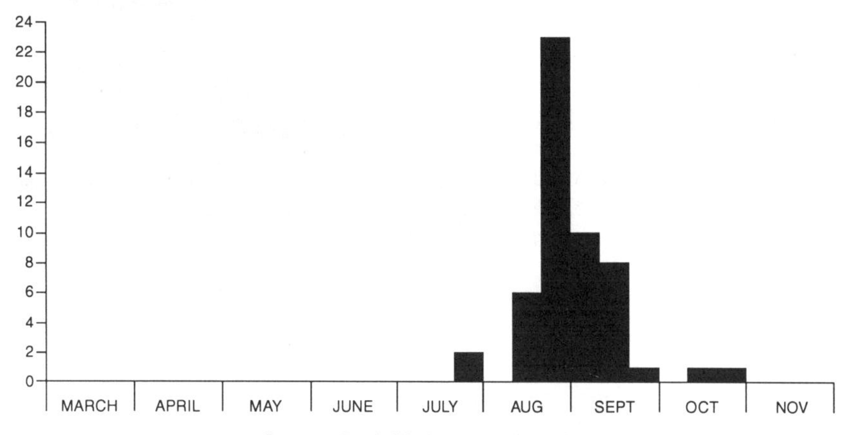


***Orange Ladybird – records of larvae***

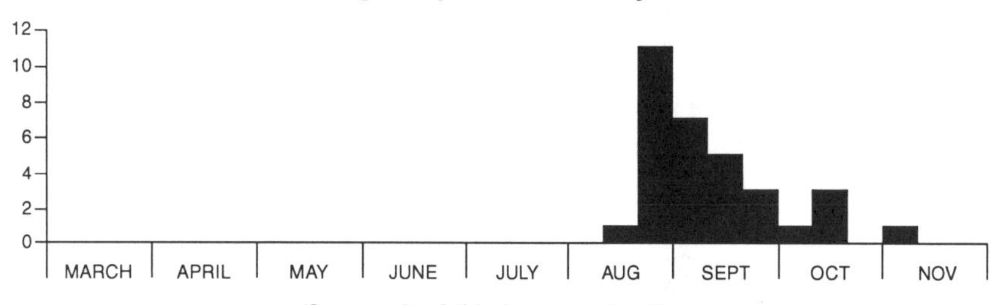


***Orange Ladybird – records of pupae***

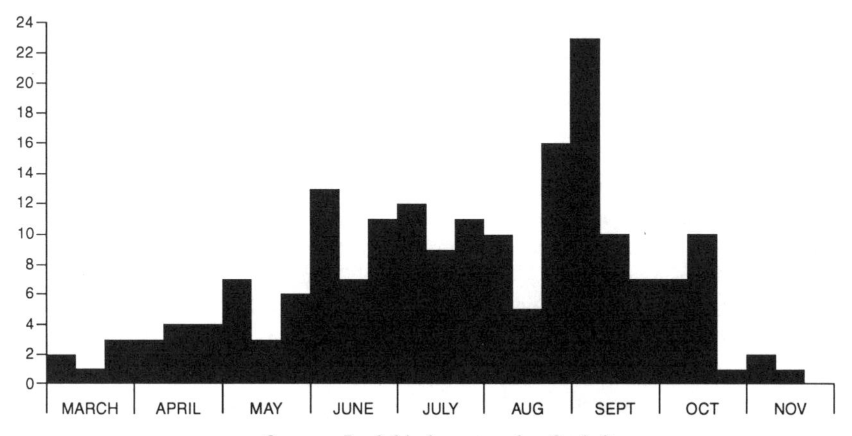


***Orange Ladybird – records of adults***

## *Psyllobora 22-punctata* (Linnaeus, 1758) PLATES 7, 15 **22-spot Ladybird**

(= *Thea 22-punctata*)

**National Status:** Common
**Number of Tetrads:** 352
**Status in Surrey:** Ubiquitous
**Habitat:** Mildewed leaves of low-growing plants

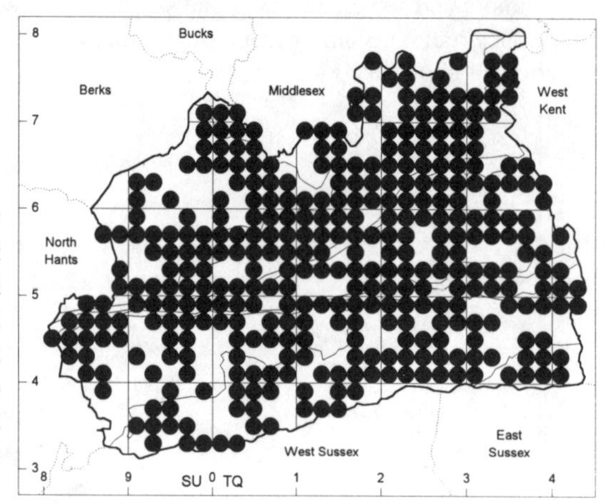


This small, bright lemon-yellow ladybird is very distinctive and not easily forgotten. It has many round black spots, either 20 or 22 in number, with no tendency to join up. It is found throughout the county and is at least as common in towns and in inner London as in the countryside.

The only species which can be confused with the 22-spot is the 14-spot in its variety with all the spots separated. The latter may be quite a bright yellow when fully mature, but of course it only has fourteen spots and those in the position of the 'anchor' are rather square. Any doubt can be removed by looking at the pronotum and the base of the wing-cases. The pronotum of the 22-spot has a semicircle of four black spots (no outer spots) surrounding a basal spot in the shape of a forward-pointing triangle, behind which the basal area of the wing-cases is wholly yellow (no black line).

This is the only species of ladybird whose colours stay the same throughout the larval, pupal and adult stages – all are yellow with black spots. The larva can only be confused with that of the Orange Ladybird. The 22-spot larva has shorter legs, its head is dark brown and the middle tubercle on each side of the first abdominal segment is entirely yellow, although all the others are black.

Both the 22-spot and Orange Ladybirds feed on mildew but are rarely found together, since they live at different levels. The Orange Ladybird feeds beneath the mildewed leaves of trees and shrubs, while the 22-spot grazes mildew off the upper surface of leaves of herbaceous plants and is especially noticeable on hogweed. Young trees growing out of grassland or heath are particularly prone to mildew and 22-spots are also often found on these, or on the new growth after coppicing, or on hedges. Unlike the Orange Ladybird, the 22-spot prefers small oaks in these situations, but is rarely found above one metre from the ground. Both species are found on dogwood, whether growing as isolated bushes or in hedges, and on sycamores where the lowest branches reach down to the ground, and these appear to be the only meeting points between the two species. We have only one record of them actually occurring together: one adult of each species under a sycamore leaf in Leatherhead on 6 October 1999.

In winter 22-spots are usually to be found in grass tufts, often around a young tree or other plant on whose mildewed leaves they have been feeding, but we have also beaten some from ivy. Our few records of mating pairs occupy an extended period: 2 May 1993, 16 May 1994, 17 & 19 May 1997, 15 June 1992, 14 July 1985 (2 pairs).

The dates of records of larvae are plotted in the chart, which is open to different interpretations. I suggest that the main breeding cycle involves larvae maturing in August and relates to the development of mildew on the leaves of hogweed and other umbellifers. A few specimens may start to breed before this on plants attacked by mildew earlier in the year, and there may occasionally be a small second generation. It may be significant that the earliest and latest records of larvae both come from the London area: 10 June 1995 at Nunhead Cemetery and 14 October 1999 at Camberwell.

The leaves of hogweed appear to be most suitable for the species in late August and September, but as they die back they lose their attraction and there is a definite switch to young trees, especially oaks, and other herbaceous plants (see charts). This is not simply a matter of the ladybird breeding on hogweed and the fresh adults moving to other plants, since larvae have also been found on young oaks on six occasions.

Out of a total of 998 specimens recorded from plants with obviously mildewed leaves and presumably feeding on them, 31% were on hogweed, 24% on young oaks and 10% on creeping thistle. The full list of host-plants for the 22-spot Ladybird reads like a complete list of the British flora, so we list only those on which at least 10 specimens

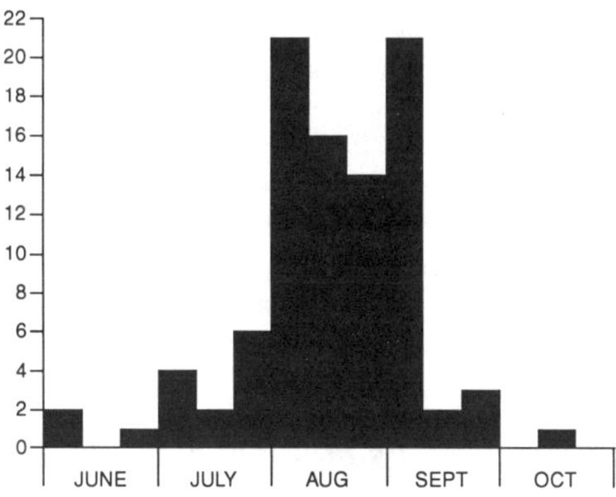


***22-spot Ladybird – number of larvae***

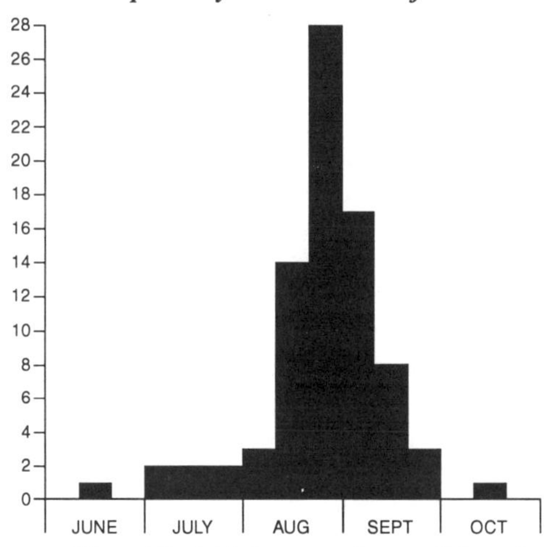


***22-spot Ladybird – records on mildewed hogweed***

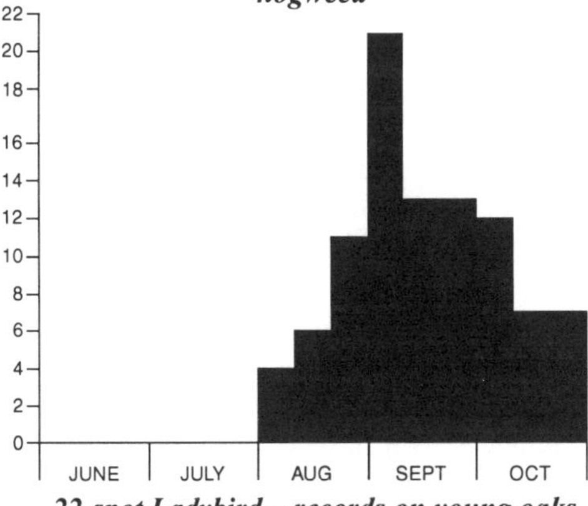


***22-spot Ladybird – records on young oaks***

(1%) were recorded. It would however be a shame not to mention such delightful obscurities with single records as small bugloss, marsh woundwort and the Duke of Argyll's tea-plant. This introduced shrub does in fact qualify for our list through five adults and eight larvae (1.3%) feeding on its mildewed leaves, even though all were on a single plant at Byfleet. Other host-plants are dogwood and mugwort (both 3%); Canadian goldenrod, young sycamores, meadow vetchling, common knapweed, spear thistle and wild parsnip (all 2%); nettle, hedge woundwort, angelica and tansy (all 1%). The 22-spot Ladybird is thus one of the few insects to be associated with the Canadian goldenrod, a garden escape that is spreading rapidly on railway banks in Surrey and also invading chalk downland.

## *Cryptolaemus montrouzieri* Mulsant, 1853

This Australian ladybird has been deliberately introduced, for the purposes of biological control, to many parts of the world where the climate is tropical or Mediterranean. In Britain it is used commercially for the control of pests under glass and in butterfly houses, where poison sprays would be injurious to the butterflies being reared. Escapes are always possible and two specimens were found out-of-doors in Yorkshire in July 1991, but it is unlikely that this species from a frost-free region would survive a winter here (Constantine & Majerus, 1994). The insect is 4 mm long and the entire upper surface is covered with short hairs. The wing-cases are black, given a greyish tint by the covering of pale grey hairs, but the head, pronotum and extreme tips of the wing-cases are dull orange. It is illustrated in black-and-white by Majerus (1994) and in Majerus & Kearns (1989). The larva is covered with thick wax.

At least two establishments in Surrey have used *Cryptolaemus montrouzieri* for pest control, the Royal Botanic Gardens at Kew and the Royal Horticultural Society's Garden at Wisley, where two colonies of the ladybird, both adults and larvae, were found outside in late September 1997 on plants of *Phormium tenax*, the New Zealand flax (Halstead, 1999). This is a huge plant with upright, fleshy, sword-shaped leaves reminiscent of its relatives *Yucca* and *Agave*. The ladybirds were feeding on a mealybug, *Balanococcus diminutus*, that has been imported into Britain on its host-plant and is now established here, being tolerant of frost. Andrew Halstead kept one colony under careful observation to see how long it would survive. The last adult was seen on 22 October shortly before the onset of air frosts, but larvae were present throughout November and up to 10 December, having survived about 15 nights of air frost with a minimum temperature of –5.7 °C. A single adult was found on a warm day in February. It is unlikely to have been a fresh escape at that season, but otherwise must have survived a further 15 nights of frost including a prolonged period of cold weather.

These observations demonstrate that the ladybird is more tolerant of frost than had previously been supposed, although the colony still did not persist through the winter of 1997/8. One wonders whether the mealybug and its ladybird predator are present in frost-free parts of south-west England such as the Scilly Isles where horticulture is the main industry and *Phormium tenax* and its smaller relative *Phormium cookianum* are persistent where planted and have become naturalised (Stace, 1997).

# SMALLER LADYBIRDS

This grouping is largely artificial since it contains species from different subfamilies of the family Coccinellidae. The separation originated with the Cambridge Ladybird Survey whose amateur observers could not necessarily be expected to recognise these small beetles as ladybirds or to collect specimens and name them.

All these species are known by scientific names rather than English names. They are usually small and often unspotted, so recognising them as members of the ladybird family is not easy without some experience. After 15 years of recording these insects, I still ocasionally take home a beetle from another family by mistake. Even when certain that one has a ladybird, it is often difficult to name it to species in the field, so a specimen should always be taken for examination under a strong lens or binocular microscope.

The key by Majerus & Kearns (1989) can be used for identifying the smaller ladybirds, although I have myself always used the key in the comprehensive and masterly paper by Pope (1973) for the genera *Scymnus* and *Nephus*, and now prefer the German-language key by Fürsch (1967) for the other genera. The key by Joy (1932) is perfectly adequate for the larger ladybirds (except for new arrivals) but is no longer satisfactory for *Scymnus* and *Nephus*. It is important to check the structural details of the underside in these two genera.

One particular tiny beetle is often mistaken for a small ladybird. It is commonly found in houses and is patterned with wavy bands formed by yellow and white scales. Its occurrence indoors is no accident, for it is a household pest. This is one of the genus *Anthrenus*, notorious for destroying natural history specimens including those collections of pinned insects that are so essential as reference material to anyone attempting their identification. In a private house it is more likely to be feeding on carpets or clothing. It is a Bad Thing and should not be tended and released like a ladybird found indoors, but firmly squashed.

## *Coccidula rufa* (Herbst, 1783)

**National Status:** Common
**Number of Tetrads:** 62
**Status in Surrey:** Local
**Habitat:** Tall vegetation in marshes or by water

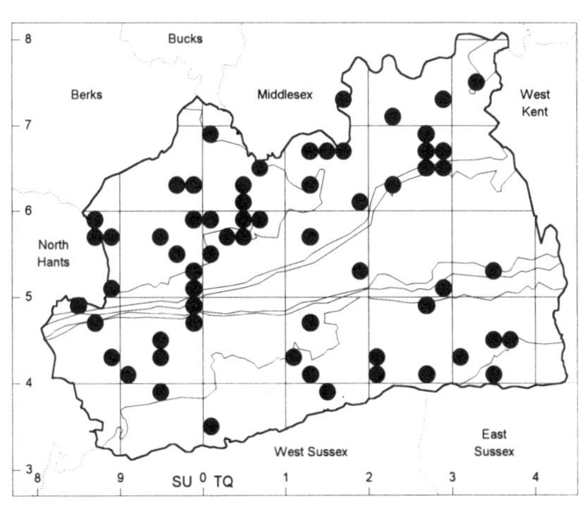


The name of Red Marsh Ladybird has been proposed for this species (e.g. Burton, 1968) and seems particularly suitable through combining its colour with its habitat. At 3 mm long it is as large as the 16-spot but a little smaller than the Water Ladybird with which it is often found. The wing-cases are elongate, distinctly reddish-brown, and the entire upper surface is covered with short hairs.

The difficulty in recognising *Coccidula rufa* is in knowing that it is a ladybird at all, since the antennae are long, almost reaching back to the base of the pronotum, and the body is flattened and much longer than broad. Many other small beetles have this shape, particularly in damp habitats, but few if any that share its habitat have the same reddish colour.

Most of our records are from ponds but the species is also found in wet meadows and by rivers and canals. The distribution map shows a line of dots along the Wey Navigation from above Guildford (SU9846) to where it joins the Thames (TQ0644). The occasional specimen turns up in dry habitats.

It is often found behind the leaf sheaths of reedmace and other waterside plants such as yellow iris in late autumn and early spring, even as late as 10 May in 1986. Other overwintering sites are tufts of grass and sedges. In spring and late summer it occurs on many wetland plants, most usually on reed sweet-grass by the Wey Navigation.

## *Coccidula scutellata* (Herbst, 1783)

**National Status:** Local
**Number of Tetrads:** 9
**Status in Surrey:** Very local
**Habitat:** On reedmace beside ponds

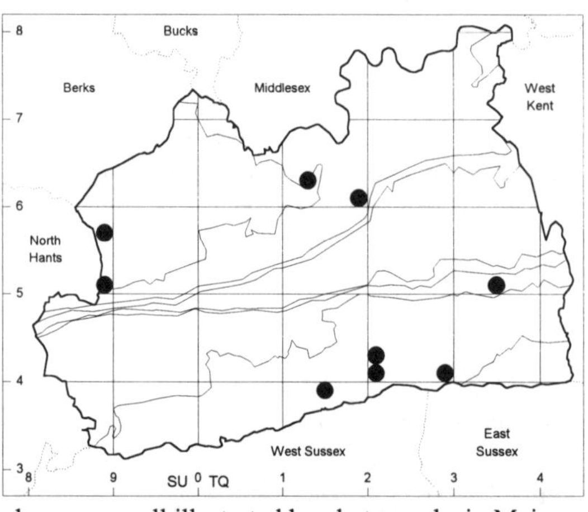


This species is very similar in size and shape to *Coccidula rufa*, but it is slightly paler and more orange-brown in colour and has a consistent pattern of five black spots on the wing-cases. It is rather difficult to separate from *C. rufa* except through these dark marks, but the male genitalia are quite distinct. The adults, larvae and pupae are all illustrated by photographs in Majerus (1994).

As far as is known, all our records come from reedmace growing beside ponds, both large and small. It occurs on the leaves in late summer as well as behind the leaf sheaths in winter. Counts of the number of ladybirds on different types of vegetation by a pond in Germany found both species of *Coccidula* most abundant on common reed (uncommon in Surrey), but 31% of *C. scutellata* were on reedmace compared with only 4% of *C. rufa* (Klausnitzer, 1997).

The distribution of *C. scutellata* in Surrey represents something of a mystery. It has not been found again at any of the sites in the south-east of the county where it was recorded in 1984/5 and 1990. The pond at Gatwick Airport still has Water Ladybird but *C. scutellata* appeared to be absent in April 2000. Reedmace is still present but now grows in permanent water adjacent to sallow bushes at the edge of the pond. In 1985 there was a large expanse of reedmace that was flooded only occasionally.

The species was found between 1987 and 1989 at both the Lesser and Great Stew Ponds on Epsom Common by parties from an FSC course led by Michael Majerus. The Great Stew Pond was restored in the late 1970's but now has a considerable growth of scrub around its edges. In 1999 there were Water Ladybirds on reedmace and *C. rufa* on rushes but I found no *C. scutellata*. Surely it cannot really have been lost from this fine site.

The rise and fall of the species at Gatwick Airport and Epsom Common suggests a link to the age of the pond, perhaps to an early stage in the succession when reedmace is dominant, but the records made by Jonty Denton in 1999 come from long-established waters such as Vann Lake as well as those of more recent origin in former gravel pits.

RECORDS: **Newdigate Brickworks** (TQ2042), 5+ on leaves of reedmace, 25.8.84 (RDH); **Green Lane Farm, Newdigate** (TQ2041), one on reedmace, between sheath of dead leaf and stem, 8.12.84 (RDH); **Gatwick Airport** (TQ2841), two behind leaf sheaths of reedmace, 20.4.85 (RDH); **Epsom Common** (TQ1860), on reedmace by ponds, 30.8.87, 28.8.88, 27.8.89 (RDH); **Bay Pond** (TQ3551), 10.8.90 (GAC); **Vann Lake** (TQ1539), 1.6.99 (JSD); **Lakeside Park, Ash** (SU8851), 1999 (JSD); **Coleford Lake, Mytchett** (SU8856), 1999 (JSD); **West End Common, Esher** (TQ1263), 27.4.99 (JSD).

## ***Rhyzobius litura*** (Fabricius, 1787) PLATE 13

**National Status:** Common
**Number of Tetrads:** 403
**Status in Surrey:** Ubiquitous
**Habitat:** Long grass and thistles

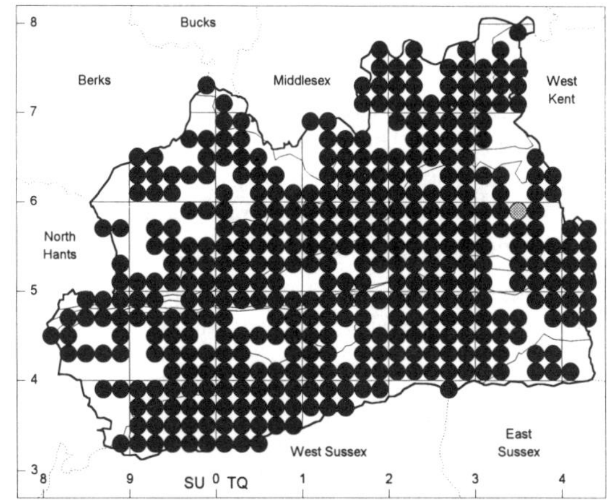


The name of Small Brown Ladybird has been proposed for this unassuming species, but the name has not caught on and the insect is perhaps not distinctive enough to deserve an English name. It is a small, oval, yellow-brown beetle whose entire upper surface is covered with short hairs. The antennae are remarkably long for a ladybird, extending back to reach the wing-cases, with the last three joints forming a narrow club. Other similar beetles of this size and colour either lack the downy hairs or have different antennae, either slender or with a broad spherical club. There is often a dark, squarish U-shaped mark on the wing-cases, while some paler and more yellowish individuals occur, particularly early in the season when they are presumably freshly-emerged; occasional examples are very dark, almost black.

Its typical habitat appears to be undisturbed grassland, especially where creeping thistle grows out of long grass. It was never found in large numbers, with the maximum at one locality being 27 on 18 September 1998 along a 300-metre length of farm track. Two more records of 10+ came from long grass around young planted trees. The few records from early in the year were from grass, nettles and buttercups. Later in the season it was found almost everywhere on creeping thistle and in grass tufts, and also on spear thistle, mugwort and fleabane; there appears to be a preference for grasses and composite plants, but we have records from grass without thistles and thistles without grass. Almost all records were made by tapping such plants sideways over a white sheet. Very few observations were made of undisturbed insects since they are so small, but one was observed on the fruiting heads and stem of hogweed among black-and-green aphids. However, larvae of this species were reported by Mills (1981) as feeding on certain aphids on knapweed, creeping thistle and cock's-foot grass, while studies on the Continent have shown that the species can feed on pollen and fungal spores as well as on aphids (Majerus, 1994).

In 1996 it was realised that this species was being found at every locality that was searched in autumn, so over the next three years an effort was made to record it systematically throughout the county. This process is not quite complete, for it is likely to occur in almost all squares. The few squares examined where it was not found were either dense woodland or extremely built-up, but even here some potential habitat was clearly present on private land such as railway banks or allotments.

The dates of records are summarised in the chart opposite which shows the number of days on which the species was recorded rather than the number of records, in order to mask the unbalancing effect of systematic recording in the autumn when many records were made each day. The pattern is consistent with overwintering adults forming a single generation each year. There are scattered records throughout the spring, mostly in May and June, although no particular search was made for it at this time. The only mating pair was observed on 4 June 1989. After a gap in July, numbers build up during August to a peak in September, with immature adults present until the middle of that month. The species is difficult to find from mid-October onwards, presumably dropping down as the weather gets colder; one was sieved from moss on 15 October 1998.

It was strangely absent from an area of apparently suitable habitat of long grass and creeping thistle by the River Thames at Petersham Meadows in October 1999. It may be significant that *Coccidula rufa* was present (six found) and the area lay in a depression that perhaps floods in winter.

It is a quirk of this survey that almost all the records for this species (93%) are my own, since ladybird enthusiasts overlook it as just another small brown beetle, while beetle specialists tend to ignore it because it is so common. Since the discovery in Britain of the closely-related species *Rhyzobius chrysomeloides*, which is only certainly separated from *litura* by dissection, I have been checking all records of *Rhyzobius* by dissecting each specimen taken. This confirms that the species so common in areas of long grass and thistles is indeed always *litura*, and incidentally shows males and females to be present in the population in equal numbers (170 males and 172 females examined).

It is interesting to examine a specimen closely under a binocular microscope or strong lens. The yellow-brown colour is not uniform but includes minute brown spots at the base of the hairs on the wing-cases. The transparent wings have dark marks at the tip and along the front edge and are folded three times, once longitudinally and twice from end to end, so that there are eight thicknesses of wing membrane packed beneath the wing-cases. The folded wings can be seen through the wing-cases as pale areas beside a central dark triangle resulting from the body colour. However, only a few specimens have fully-developed wings, while most have their wings reduced to vestigial strips and so are quite incapable of flight. From insects taken from many parts of Surrey during 1999, 5 males out of 57 (9%) and 8 females out of 47 (17%) were fully-winged. It has been known for almost 170 years that only a small proportion of British specimens have fully-developed wings (Pope, 1977; Hammond, 1985).

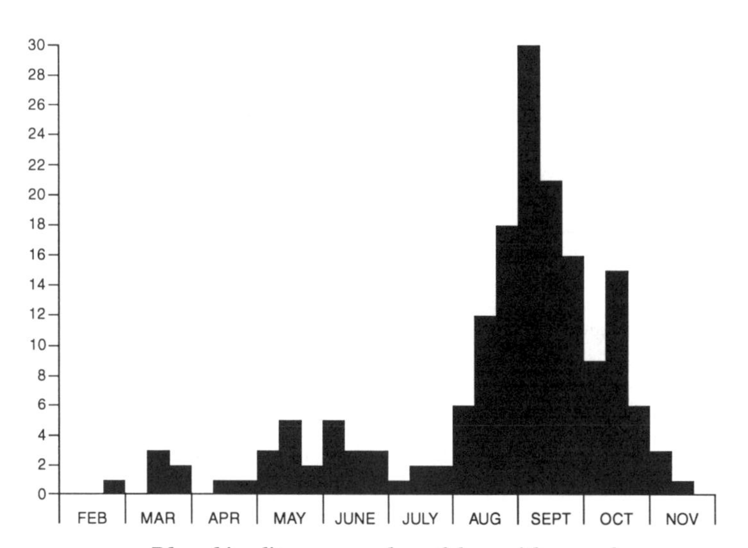


***Rhyzobius litura* – number of days with records**

## *Rhyzobius chrysomeloides* (Herbst, 1792)

**National Status:** Recent colonist
**Number of Tetrads:** 3
**Status in Surrey:** Established in at least one area
**Habitat:** Ivy, hawthorn and small pine trees

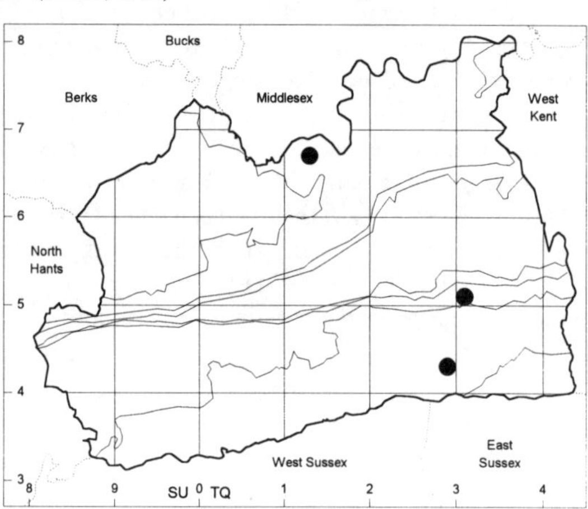


This small brown beetle is very similar to *Rhyzobius litura* and is considered to be more typical of central and eastern Europe while our common *R. litura* is found predominantly in western Europe (Klausnitzer, 1997). It was thus a major surprise when a single specimen of *R. chrysomeloides* was found on a small pine tree on a motorway bank near Nutfield in east Surrey in 1996 (Hawkins, *in press*). It was even more surprising when a thriving colony was discovered in the same Molesey garden in which the Bryony Ladybird had just been found for the first time in Britain (Menzies, 1999).

The new species, *R. chrysomeloides*, is distinguished externally from the common *R. litura* by minute structural details: the sides of the pronotum are more rounded and less tapering at the base, while the prosternal keels (between fore coxae) have a parallel-sided central section. These differences are slight and the characters vary between individual specimens and particularly between the sexes, so for certain identification, dissection of specimens is necessary. The male genitalia of the two species are very distinct and the female genital plates also differ slightly in shape and can be distinguished by careful comparison of specimens. At first glance, *R. chrysomeloides* appears to be a paler insect with more extensive black markings on the wing-cases, but these superficial characters are very variable.

The motorway bank near Nutfield, where the species was first found in Britain, has since been searched without success (RGB). In contrast, searching for the species around Molesey has yielded more and more specimens throughout 1998 and 1999, all beaten from ivy or from old hawthorn trees. The species is evidently well-established in this local area of Surrey (ISM).

The species was also found in Horley in April 2000, at a locality very like the Molesey one, with ivy growing over the fence of an ordinary suburban garden beside a hawthorn hedge. One would not normally think of searching for rare beetles in such places, and it is probable that *R. chrysomeloides* is already widespread in Surrey and possibly over much of southern England, for it has now been recorded from Eastbourne in Sussex (Hodge, *in press*).

These first records from Britain all refer to pine trees, ivy and hawthorn as the host-plants for the species, and on the Continent it is also most frequently found on pine trees and bushes, especially near water (Fürsch, 1967). Since it lives in a different habitat from *R. litura*, which is usually found on grass and thistles, it presumably also feeds on different things. The three Horley specimens comprised a fully-winged female and two males with vestigial wings – another aspect worthy of investigation.

RECORDS: **Nutfield** (TQ3151), one beaten from young pine tree at foot of motorway embankment, 29.4.96 (RDH); **West Molesey** (TQ1367), beaten from ivy and hawthorn at three separate sites, 1.11.97 (several specimens) (ISM), 4.11.97 (6), 8.11.97 (frequent), 6.6.98 (several), 29.6.98 (6), 6.7.98 (several), 12.7.98 (several), 15.7.98 (several) (ISM, JAO, RGB & others); found in similar numbers during 1999; **Horley** (TQ2843), tapped 3 from ivy growing over garden fence, 29.4.00 (RDH).

## *Rhyzobius lophanthae* (Blaisdell, 1892)

(= *Lindorus lophanthae*)

This originally Australian species was found for the first time in this country on 1 April 1999 by Derek Coleman (Booth, *in press*). Because of the date of the record, it was not at first believed.

The single specimen was crawling up the trunk of an ash tree in Morden Park (TQ244672). It was a small black ladybird with the long antennae typical of *Rhyzobius*. Derek is familiar with *Rhyzobius litura* and some species of *Scymnus*, and recognised it as an unusual coccinellid. He took it to Roger Booth who named it as *Rhyzobius lophanthae*, a native of eastern Australia. The species was introduced to Italy in 1908 to control scale-insects, particularly the hard scales of the subfamily Diaspidinae. After further introductions, it has since spread widely around the Mediterranean basin and has been introduced to several other parts of the world (Greathead, 1973).

The specimen possibly came from a nearby tree nursery. Further searches of the area have so far been unproductive. Since the species normally lives in frost-free regions of the world, it is unlikely to withstand cold winters and become a permanent member of our fauna, although there have been attempts in Europe to develop a strain of the species more resistant to cold (Klausnitzer, 1997).

## *Scymnus femoralis* (Gyllenhal, 1827)

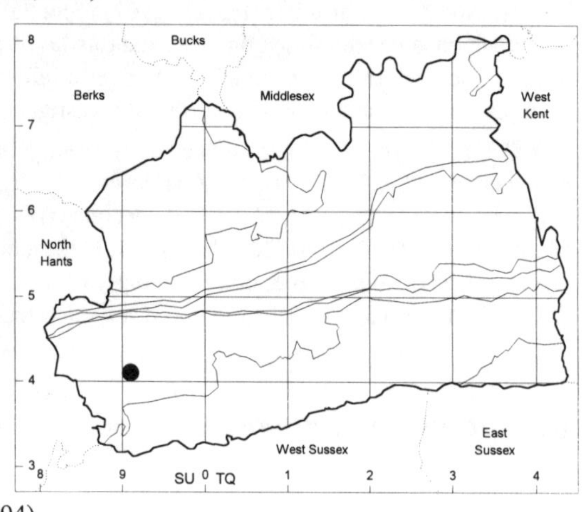


**National Status:** Notable (NB)
**Number of Tetrads:** 1
**Status in Surrey:** Rare
**Habitat:** Dry places

This seems to be the rarest in Surrey of the *Scymnus* species living close to the ground, with just one record from a heathland National Nature Reserve during the 20 years of the survey. In Britain as a whole it is a local species found in low-growing vegetation on well-drained soils, particularly chalk downland, sand dunes and heathland (Majerus, 1994).

This species and the three following belong to the typical subgenus of *Scymnus* which has the postcoxal lines incomplete (J-shaped) as well as having two low ridges (carinae or keels) on the prosternum. *S. femoralis* is black with pale antennae, mouthparts and legs, except for the basal half of the femora, and needs careful separation from *S. schmidti*. When specimens are compared, the body of *S. femoralis* is usually distinctly more convex and that of *S. schmidti* is flatter.

The following specimens from museum collections appear to be mainly from chalk downland and some heathland sites: Guildford (G.C. Champion); Ewhurst (Tomlin); Box Hill (Donisthorpe); Mickleham (Power); Kingswood (Saunders); Esher, Merstham, Reigate and Norwood (unknown collectors) (all verified by R.D. Pope). The inescapable conclusion that the species has become rarer is perhaps merely a reflection of how much of our downland and heathland has become wooded where it is no longer grazed.

RECORDS: **Thursley Common** (SU9040), 17.8.97, (JSD) (*The Coleopterist* **8**: 21).

## *Scymnus frontalis* (Fabricius, 1787) PLATE 13

**National Status:** Common
**Number of Tetrads:** 18
**Status in Surrey:** Local
**Habitat:** Bare ground with sparse vegetation

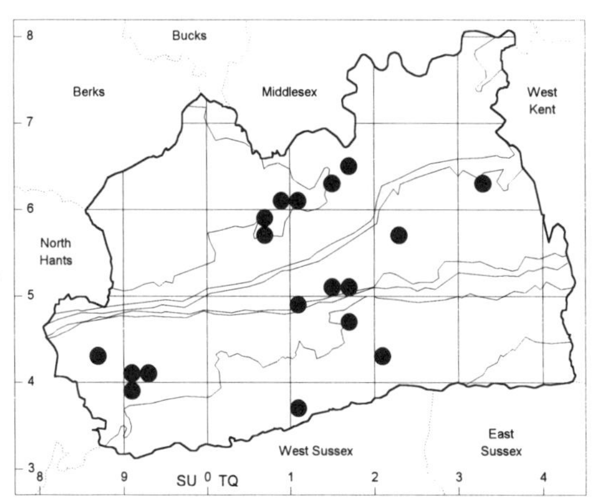


This is the largest species among *Scymnus* and the related genus *Nephus*, and the most common of those found close to the ground. It is black with a pair of elongate red spots positioned towards the front of the wing-cases. We have found it in four distinct habitats on different types of soil.

On the clay soils in the south of the county, it is a characteristic species of the pits created by excavating the clay to make bricks. The ladybird occurs at three of these sites and always on bare south-facing slopes. Several larvae, with their white spiky outgrowths, were found by Gail Jeffcoate in the clay pit on Inholms Lane at North Holmwood on 13 July 1998, with another 8 to 10 larvae at the former brickworks in Somersbury Wood on 3 August. I succeeded in rearing one larva to the adult stage and confirmed the species for the other site by revisiting it.

At the former brickworks at Newdigate on 17 July 1984, several larvae were on common knapweed feeding on some black aphids resembling the illustration of *Dactynotus jaceae* in Rotheray (1989). These aphids had defeated a 7-spot Ladybird which ate two or three but then retired from the fray, wiping its face, following concerted and agitated movement by the aphids; the ladybird remained motionless for several minutes and rejected the aphids when it encountered them again. However, the presence among them of the larvae of the much smaller ladybird did not produce this defensive reaction. This aphid cannot be a regular food, since I have not seen *S. frontalis* on knapweed again. At the same site on 3 July 1998, ten larvae were counted in a small area of more usual habitat, walking on bare clay and over short vegetation grazed by Canada geese.

This close grouping of the dates of four records of larvae, from 3 July to 3 August, suggests that the species has a single generation a year.

Several records come from chalk downland, including one specimen found among moss in mid-April, presumably having overwintered there. On the heathlands of west Surrey, Jonty Denton has found it repeatedly by sweeping sheep's sorrel on Thursley, Witley and Hankley Commons.

In 1999, which must have been a good year for this species even though the populations of some other ladybirds had crashed, it appeared among coarse vegetation at several roadside sites in early September. The habitat was distinct from its other localities, since the ladybirds were tapped from grass tufts, nettles and young trees up to a metre high.

## *Scymnus nigrinus* Kugelann, 1794

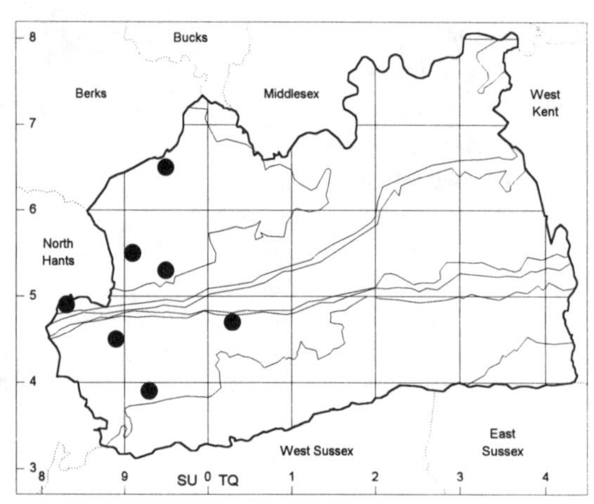


**National Status:** Local
**Number of Tetrads:** 7
**Status in Surrey:** Very local
**Habitat:** Young pine trees

A small black ladybird beaten from Scots pine is likely to be this species. It is about 2.6 mm long and entirely black except for the tarsi which vary from pale to very dark brown. Our few records all come from the heaths of west Surrey where pines are invading the heathland. All but one of the records were from young pine trees (1 to 4 metres high), while the other was from a sallow growing among pines. This species thus has the most restricted distribution of any of the pine specialists.

At Chobham Common on 25 April 1987 I noted "about six adults seen on twigs of little pine with much white fluff and tiny fawn objects inside it [eggs of adelgids]; the ladybirds run rapidly, fly easily and drop off when disturbed; they inspect the white fluff but do not stop; instead they appear to feed under scales of bark." Although this observation lacks precise information as to the prey of the ladybird, it might be worth following up. A necessary first step would be to distinguish the different species of tiny insect, such as aphids and coccids, living on the pines.

The species is likely to be not quite as rare as the map suggests, since some of our recorders did not tackle the smaller ladybirds. It probably still exists at the following sites from which it was recorded in the 1960's: Thursley Common (post-1960, K.C. Side), Horsell Common (20.8.67, B. Levey) and Oxshott Heath (A.A. Allen) (all verified by R.D. Pope).

RECORDS: **Oxshott Heath**, beat one from young pines, 31.8.60 (AAA); **Blackheath, near Guildford** (TQ0346), on seedling pines, 19.8.84 (RDH); **Chobham Common** (SU9564), six on little pine (observations in text), 25.4.87 (RDH); **Bricksbury Hill** (SU8349), beaten from sallow, 4.9.88 (RDH); **Spur Hill** (SU9054), beaten from pine, 2.7.89 (GAC); **Mare Hill Common** (SU9239), beaten from pine, 5.5.96 (RDH); **Crooksbury Common** (SU8845), beaten from young pine, 30.8.97 (RDH); **Cobbett Hill Common** (SU9453), beat two from pines, 9.5.98 (RDH).

## *Scymnus schmidti* Fürsch, 1967

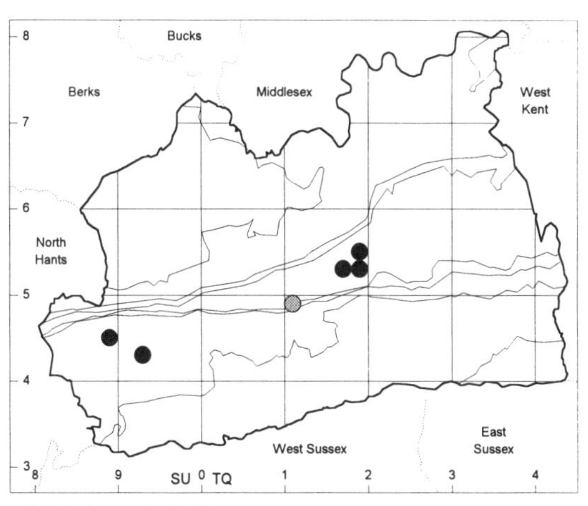


**National Status:** Notable (NB)
**Number of Tetrads:** 5
**Status in Surrey:** Rare
**Habitat:** Dry places

This is another black species but with pale tibiae and so very similar to *S. femoralis*. It is slightly larger and more elongate and has a distinctive feature on the underside: a groove down the centre of the metasternum. Nationally it is a very local species of low-growing vegetation on well-drained soils, occurring on sand dunes, chalk downland and heathland (Majerus, 1994).

Most of our few Surrey records come from the North Downs around Dorking, but there is one record from heathland in west Surrey and a slightly anomalous one from wetland by the River Wey at Eashing. This was found in or near flood debris, so had probably been washed down from the vast expanse of dry heathland drained by this river and its tributaries.

The historical records from museum collections seem also to be mostly from chalk downland with just the first from heathland: Chobham (Donisthorpe), Box Hill (T.H. Beare), Mickleham (Power, Sharp), Kingswood (Power), Coulsdon (Henderson) (all verified by R.D. Pope). The distribution and status of the species do not therefore seem to be greatly altered.

RECORDS: **White Downs** (TQ14), 1976 (PJH); **Box Hill** (TQ1852), 19.4.85 (PJH); (TQ1752), swept, 16.4.87 (JAO); **Headley Warren** (TQ1954), sieving grass and moss from short turf, 16.4.94 (RGB); **Eashing Valley** (SU9343), 5.6.95; **Crooksbury Common** (SU8945), 4.7.95 (both JSD, *The Coleopterist* **8**: 21).

## *Scymnus auritus* Thunberg, 1795 PLATE 13

**National Status:** Local
**Number of Tetrads:** 23
**Status in Surrey:** Local
**Habitat:** Leaves of oak trees

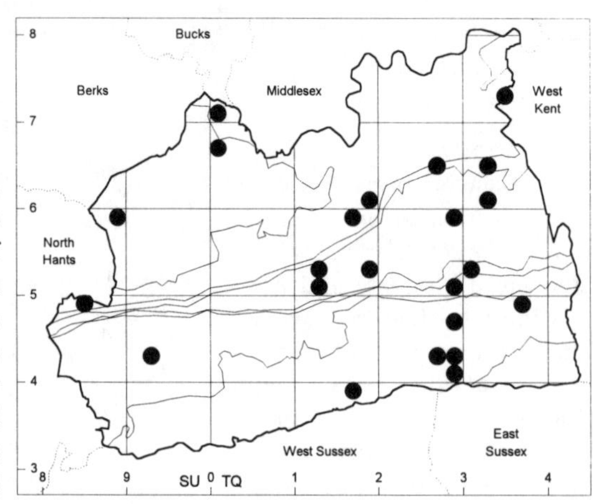


The female of this species is almost wholly black but the male has its head and the front edges of its pronotum distinctively red. On the underside, the tip of the abdomen and much of the legs are also coloured red. There are many old records from Surrey and between 1963 and 1968 it was found in large numbers around Kew (Eastop and Pope, 1966, 1969). It was very closely associated with oak leaves attacked by *Phylloxera glabra.* The predator followed the cycle of its prey to be common in 1963 and 1964, absent in 1965 and 1966, and common again in 1968.

*Phylloxera* is related to aphids although now placed in a separate family, the Phylloxeridae. Unlike most aphids it has no cornicles (the pair of upright 'tails' on the abdomen which give off a defensive spray). The female feeds at one spot beneath an oak leaf and eventually becomes surrounded by a circle of eggs. Although *Phylloxera* causes yellow spots on oak leaves, not all such spots are caused by *Phylloxera. P. glabra* is the only phylloxerid feeding on oak leaves in this country, although another species, *Moritziella corticalis*, may occur on the twigs (Barson & Carter, 1972).

In 1985 I found *Scymnus auritus* at Croydon, Capel and four sites in Horley, and was able to observe its development. On 15 & 16 May single adult ladybirds were found under oak leaves along with adults and young of a brown hump-backed aphid with short antennae and no cornicles, not positively identified as *Phylloxera.* Taken into captivity, the *Scymnus* ate a young aphid. By 16 July there were many larvae beneath the oak leaves near some round orange aphids each surrounded by a circle of eggs, matching the appearance and behaviour of *Phylloxera.* The coccinellid larvae came in three sizes and were covered with waxy outgrowths giving the appearance of small white hedgehogs, except when newly-moulted. Two were taken into captivity and emerged as adults by 12 August. At Capel on 1 August, a *Scymnus* larva was again associated with *Phylloxera*, but by now the mother aphids had died and fallen off leaving the characteristic yellow spots, while their eggs had hatched and aphid nymphs were wandering over the leaves.

Observations in 1986 followed the same pattern but adults of the new generation had emerged by 31 July. In 1989 oak trees in Horley and Salfords were infested with *Phylloxera* and four *Scymnus* larvae found on 24 & 26 June were all reared to adult by 11 July. Another from Riddlesdown pupated by 11 July and emerged by 28 July. Remarkably, two further larvae were found on the trunk of a lime tree in Horley on 24 June. This tree is in a field close to an oak with touching foliage. Both these larvae were also reared to adult to confirm the species. A few larvae were also found in 1990.

Since then there has been a remarkable paucity of records of *auritus* with none at all of its larvae or of *Phylloxera*, perhaps because of fluctuations in numbers or maybe just through lack of searching. Adult *auritus* have been beaten from oak at about a dozen sites, with two from sycamore and one from lime. The resulting distribution map looks very sparse for a species that may well be common and widespread in favourable years.

## *Scymnus haemorrhoidalis* Herbst, 1797

**National Status:** Common
**Number of Tetrads:** 7
**Status in Surrey:** Very local
**Habitat:** Undisturbed grassland

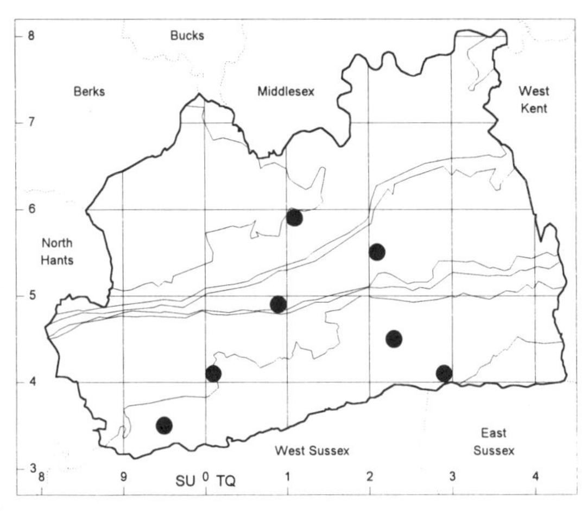


The distinctive feature of this species is a broad red tip to the wing-cases (one-fifth to one-third of their length), a character shared by other tiny beetles found in similar situations such as phalacrids. Once certain that one has a coccinellid, then it is advisable to check that the postcoxal lines are complete (U-shaped). This is the defining character of subgenus *Pullus* to which *auritus*, *haemorrhoidalis* and the two following species belong. The head, front angles of the pronotum and much of the legs are also the same dull-red colour, more extensively so in the males.

Although this is nationally considered to be a common insect, I must confess to a feeling of great excitement whenever I find one, since this has been an infrequent event. This apparent rarity is probably an illusion, since most of the sites of its capture have been rather ordinary, such as tufts of dense, coarse grass on roadside verges and the like. One feature in common to most sites has been that the grass was growing on a bank or overhang, so the sheet could be inserted right at the base of the grass tuft before tapping the grass towards it. This suggests that the species is found so rarely because it lives close to the ground. One or two other records come from higher up the grass tufts or on thistles.

RECORDS: **Bookham Common**, in vegetable refuse, 31.3.48 (AAA); **Cobham**, 2 in flood debris from River Mole, 10.82 (JAO & AAA); **Chiddingfold** (SU9535), tapped from cock's-foot tuft in neglected field, 1.9.97 (RDH); **Norwood Place Farm** (TQ2344), tapped from grass and flowers overhanging roadside bank, 2.9.98 (RDH); **Selhurst Common** (TQ0141), tapped from cock's-foot tuft at roadside, 6.9.99 (RDH); **by Great Hurst Wood** (TQ2155), tapped from creeping thistle mixed with hairy St John's-wort at field edge, 4.10.99 (RDH); **Horley** (TQ2941), tapped from grass tuft by motorway link road, 22.4.00 (RDH); **Colekitchen Down** (TQ0848), by vacuum sampler, 31.5.00 (JSD).

## *Scymnus limbatus* Stephens, 1832

**National Status:** Notable (NB)
**Number of Tetrads:** 2
**Status in Surrey:** Rare and elusive
**Habitat:** Willow and poplar trees

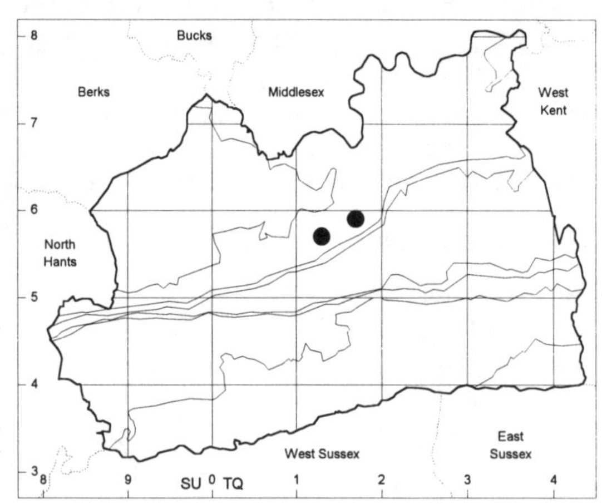


This small, neat species is apparently one of our rarest coccinellids although its habitat is a very common one. It is found on willow, sallow and poplar trees in marshy places (Majerus, 1994). What it feeds upon has not yet been determined with any certainty.

The wing-cases are mostly chestnut-brown shading into black beside the scutellum and around their edges. The insect needs very careful separation from the common *S. suturalis*, although that species lives in a different habitat, on conifers. The most useful character is that suggested by Allen (1953). The hairs in the centre of the wing-cases (at about two-fifths of their length) are directed inwards, almost at right-angles to the suture, whereas the corresponding hairs on *S. suturalis* lie parallel to the suture. On both species the hairs are directed outwards near the base and towards the tip of the wing-cases. The hair-pattern of *S. limbatus* is illustrated by Pope (1973), whereas diagram II.25 in Majerus & Kearns (1989) does not show the inward-pointing hairs adequately and appears to be closer to typical *S. suturalis* than to *S. limbatus*. If specimens are compared, the hairs on *S. limbatus* are seen to lie almost flat whereas those on *S. suturalis* include many sub-erect hairs.

Since the present species is rather similar in size and pattern to *Nephus redtenbacheri*, the underside features should be checked (postcoxal lines complete in *S. limbatus*). Differences in the colour of the legs may not be completely reliable. The legs of *N. redtenbacheri* are pale while those of *S. limbatus* have at least the femora black, according to some authors, but this is not at all obvious on the one specimen I have seen (preserved in alcohol).

We have recorded it only from Ashtead and Bookham Commons in the period of the survey, but there are older specimens in museum collections from other parts of Surrey, as follows: Woking and Walton-on-Thames (G.C. Champion); Putney and Wimbledon (Newberry); Shirley and Norwood (Power) (all verified by R.D. Pope). Although the species seems to have become rarer, it is so elusive that it may well still occur in some of these places.

RECORDS: **Wimbledon Common**, on goat willow, 6.5.50, A.A. Allen (1953, *Entomologist's Mon. Mag.* **89**: 283); **Bookham Common** (TQ1256), beaten from row of aspens six metres high, 31.5.92 (RGB); **Ashtead Common** (TQ1759), in Malaise trap, 2.7-21.8.96 (JPB per GAC, det. RDH).

## *Scymnus suturalis* Thunberg, 1795

**National Status:** Common
**Number of Tetrads:** 47
**Status in Surrey:** Local
**Habitat:** Restricted to Scots pine

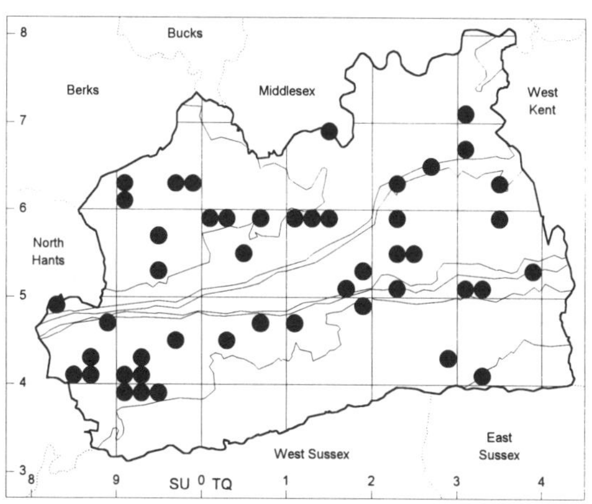


This is a small, pale brown, downy beetle, often with a vague dark mark around the scutellum. It is found almost exclusively on Scots pine and is more likely to be confused with other small pine-inhabiting beetles of similar size and colour than with related coccinellids of very different habitats. It occurs on pine trees on the heaths of west Surrey and also on trees planted in parks, on golf courses and by roadsides. The map shows an apparent absence from some areas where other pine specialists were recorded (e.g. SU83/85, TQ06). This absence may not be real since some recorders only looked for the larger ladybirds while others found it inadvisable to use the beating tray, so necessary to find such a tiny species, at certain sites such as the edge of private gardens.

Out of 125 specimens of known origin reported to this survey, 114 or 91% were found on Scots pine. Another was beaten from dense ivy on a pine trunk in October, presumably overwintering there. Two records from Douglas fir and one from larch suggest the possibility that it may also breed on these other conifers. Records from birch (well away from pine) and sycamore in October are considered as strays.

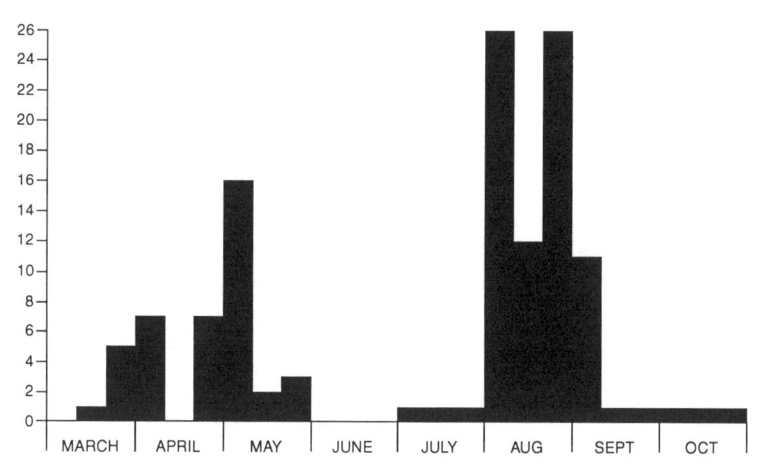


***Scymnus suturalis* – number of adults**

The dates of our records fit well with a life-history of a single generation a year with overwintering adults. These were found in April and May, or even in late March in the early seasons of 1990 and 1997. No adults at all were recorded in June and very few in July, but then the species occurred in large numbers throughout August and in the early part of September, with scattered records continuing into October.

It was surprising that the larvae of such a frequent species had not been recorded at all during the survey, so I examined some pine branches in a Horley garden in June 2000 and found that the explanation lay in the camouflage of both larvae and pupae. The slow-moving larvae with their white waxy outgrowths were present on small one-year-old shoots of pine. They were very difficult to detect among the masses of white waxy wool which occurred at intervals along the shoots and appeared to be covering the eggs of an adelgid. The larva was pale creamy-orange beneath the white wax and turned into a pupa of similar colour just over 2 mm long and covered with whitish hairs. It protruded from the remains of its wax-covered larval skin but was now the same size and colour as the spent male catkins abounding on the pine. Some mites and other aphids were present on the pine shoots as well as the adelgids, but the principal food of *S. suturalis* is given as *Pineus pini* by Klausnitzer (1997). This is the common adelgid on pines.

## *Nephus quadrimaculatus* (Herbst, 1783) PLATE 13

**National Status:** Vulnerable (RDB2)

**Number of Tetrads:** 32

**Status in Surrey:** Local but increasing

**Habitat:** Ivy on trees and walls

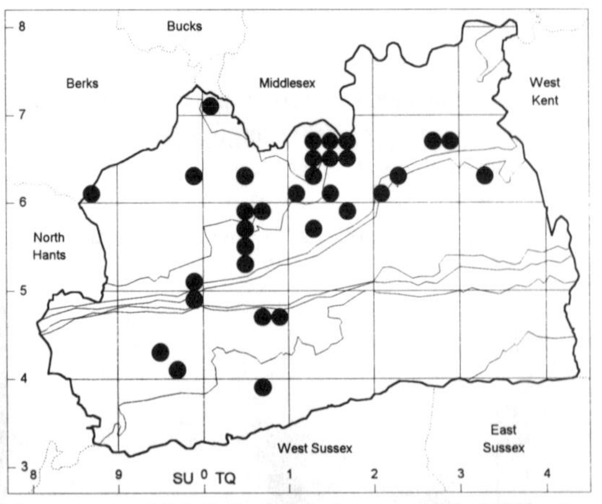


The genus *Nephus* is characterised by the absence of prosternal keels, three-segmented tarsi and post-coxal lines in the shape of a very shallow 'J', almost an 'L'. I find the tarsal character too difficult to check on each specimen, since a high magnification is needed to see the difference between a third tarsal segment which is invisible because it is hidden by another segment, and one which is invisible because it is not there.

As its name suggests, the present species has four red spots on its black wing-cases, the oval front spots pointing diagonally outwards towards the shoulders. It has always been considered a rare species in Britain with most records coming from Suffolk, in the 1890's, 1930's and 1981 when it was found in profusion on ivy by Howard Mendel (Pope, 1987). More recently it has also been recorded from two localities in West Kent (Owen, 1993; Heal, 1994), so it was not a complete surprise to find one in Surrey in 1993.

This first county record was a single specimen beaten from an isolated oak tree along with several *Scymnus auritus*. This gave no clue to its habits but subsequently it was found on ivy, initially by Andrew Halstead and then by other recorders. Over the last few years it has become rather common on ivy, whether climbing up tall trees or growing over garden walls. It is found on ivy during the summer months, so this is not merely an overwintering site. In fact it becomes harder to find the ladybird on ivy as the autumn progresses. The few records from October were obtained by brushing a stick over ivy leaves close to a tree trunk.

The distribution map indicates that it is common in the north and west of the county but absent from the south-east, where I have searched for it unsuccessfully. A record from SU84 would help to confirm this pattern and Jonty Denton has found it at Rowhill Copse in this grid square, but the exact locality was a few steps into Hampshire so the record is not shown on the map. When considered with the first record from Egham, the map suggests a rapid spread from the north-west.

Now that the species is common and its habitat is known, it should not be too difficult to find its larvae on the ivy and determine what they are eating. Records from the Continent give the mealybug *Phenacoccus aceris* as one prey species (Horion, 1961). This is a small mobile coccid living on gorse and certain deciduous trees but not normally on ivy, but there are one or two other scale-insects on ivy that might possibly be the prey (Newstead, 1903).

INITIAL RECORDS: **Egham** (TQ0170), one beaten from oak, 5.9.93 (RDH) (*Br. J. Ent. Nat. Hist.* **7**: 170); **Wisley** (TQ0659), one swept from ivy on oak tree by river, 16.9.97 (AJH) (*Br. J. Ent. Nat. Hist.* **11**: 104); **Hackbridge** (TQ2866), several by beating ivy-covered garden fence, 24.9.97 (RGB) (*Br. J. Ent. Nat. Hist.* **12**: 171); **West Molesey** (TQ1367), beaten from ivy over hospital wall and old hawthorns, 8.11.97 (many), 23.12.97 (1) (ISM with others); 1998-2000, many records.

## *Nephus redtenbacheri* (Mulsant, 1846)

**National Status:** Common
**Number of Tetrads:** 8
**Status in Surrey:** Very local
**Habitat:** Undisturbed grassland

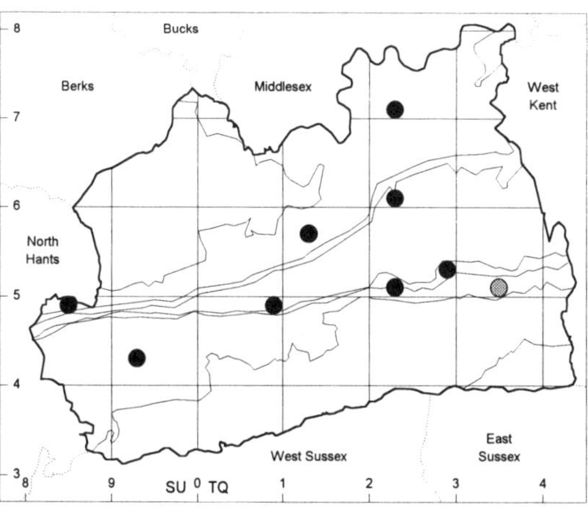


This is a widespread species in Britain, inhabiting low-growing vegetation on coastal sand-dunes and at many inland localities including boggy areas (Pope, 1973). Each of the wing-cases has a large red spot which, unlike *Scymnus frontalis*, is elongate, positioned centrally or towards the rear, and often extends over most of the wing-case. No specific prey appears to have been recorded but other members of the genus *Nephus* feed on scale-insects, particularly the mobile ones known as mealybugs.

Some of our records for this small species are from coarse grass on roadside verges or in neglected grassland, in similar sites to those harbouring *Scymnus haemorrhoidalis*, but others show a liking for grassland on sandy or acid soils and two are from chalk downland. Older records from museums and private collections show a similar diversity of habitat: Puttenham Common (post-1960, K.C. Side); Horsell, Claygate, Esher and Caterham (unknown collectors); Box Hill (T.F. Marriner); Bookham (Henderson); Buckland Hills (Power); Coulsdon (H. Britten) (all verified by R.D. Pope).

RECORDS: **Oxshott Heath or Esher Common**, 9.9.56, F.D. Buck (*Proc. S. Lond. Ent. Nat. Hist. Soc.* **1956**: 86); **Godstone** (TQ3451), 5.77 (PJH); **Merstham** (TQ2953), in clump of false oat-grass in neglected field, 1.9.90 (RDH); **Bookham Common** (TQ1255/56), 1992 (RGB); **Wimbledon Common** (TQ2371), on water plants, 21.9.95 (MVLB); **Howell Hill, Cheam** (TQ2361), grubbing at base of plants, 1996 (RGB); **Farnham Park** (SU8448), 1997 (JSD); **Bagmoor Common** (SU9242), in flight over grass and heather, 15.5.98 (RDH); **Buckland** (TQ2251), tapped from grass tufts mixed with creeping thistle, 31.10.99 (RDH); **Colekitchen Down** (TQ0848), by vacuum sampler, 31.5.00 (JSD).

## *Stethorus punctillum* (Weise, 1891)

**National Status:** Local
**Number of Tetrads:** 9
**Status in Surrey:** Very local
**Habitat:** Various

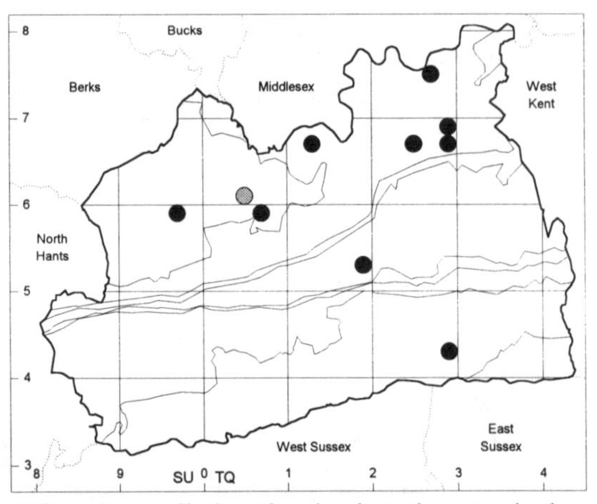


This most minute of ladybirds is never more than 1.5 mm long and entirely black apart from the antennae, mouthparts and at least the tibiae of the legs, which are all yellow. It is frequently overlooked as just another tiny beetle but, once recognised as a coccinellid, it is not difficult to separate it from those species of *Scymnus* that are also wholly or mostly black. Apart from its small size, the simplest character is that, towards the tip of the wing-cases, the hairs alongside the suture lie parallel to it rather than diverging at an angle as they do in *Scymnus*.

The other unusual feature of *Stethorus* is its prey. It is the only one of our species that is a specialist feeder on mites (Acari), although it can also eat small aphids. Some of these mites are notorious pests, so it is a very useful insect. Continental literature suggests that it is most easily found by beating shrubs and deciduous trees, and states boldly that it is especially frequent under the leaves of lime trees where it pursues the mite *Tetranychus* (Fürsch, 1967), but there we have only found aphids and their common predators such as the 10-spot Ladybird.

Our records come from diverse habitats. Andrew Halstead records it as feeding on the mite *Tetranychus urticae* in a private garden at Knaphill near Woking, and probably also on red spider mite at the Wisley garden of the Royal Horticultural Society. Others were found on hollyhock, feeding on mites, and the related common mallow, apparently eating some small white eggs – perhaps these were also mites? At the nature study centre on Wandsworth Common, several were flying rapidly among rather dried-up garden plants in a hot spell in 1995. Others were beaten from oak trees and ivy. Its small size, and the lack of a regular habitat in which to make a systematic search for it, mean that the recorded distribution probably falls far short of its true abundance.

RECORDS: **Wisley, RHS Garden** (TQ0658), 9.7.74, adult and pupal cases among red spider mites, 12.10.89 (AJH); **West Byfleet** (TQ0460), 28.6.78 (AJH); **Horley** (TQ2942), under leaf of common mallow at roadside, apparently eating some small white eggs, 4.7.85 (RDH); **Knaphill** (SU9658), feeding on *Tetranychus urticae* on garden plants, 29.8.90 (AJH); **Mitcham Common** (TQ2868), beat two from lower branches of oak, 7.10.90 (RDH); **Hackbridge** (TQ2866), feeding on plant mites on hollyhock, 15.9.91 (RGB); **Beddington Sewage Farm** (TQ2966), 4.5.92, 11.6.94 (RGB); **Wandsworth Common** (TQ2774), three flying about among dead vegetation in garden of nature study centre, 26.8.95 (RDH); **Box Hill, Juniper Bottom** (TQ1852), 7.4.95 (RGB); **Molesey** (TQ1367), beaten from ivy in private garden, 6.12.97, 12.10.98 (ISM) (*Br. J. Ent. Nat. Hist.* **12**: 176); **Morden Park** (TQ2467), 1.5.99 (RGB).

## *Clitostethus arcuatus* (Rossi, 1794)

**National Status:** Endangered (RDB1)

**Number of Tetrads:** 2

**Status in Surrey:** Rare

**Habitat:** Ivy-clad oak trunks

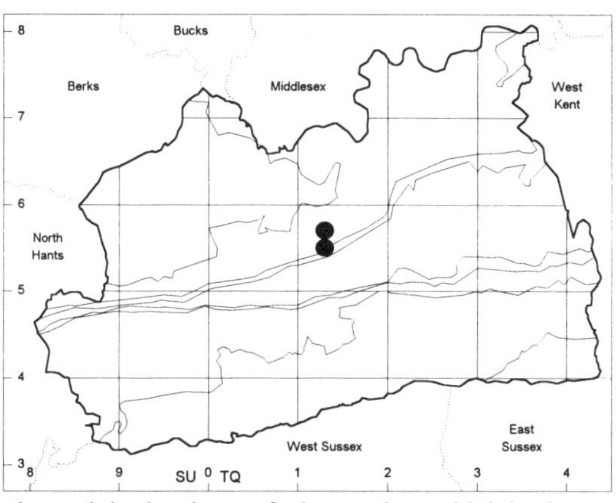


This minute species has a most attractive and unusual pattern which can be recognised on sight by anyone lucky enough to find it. It is only 1.2 to 1.5 mm long and usually dark brown in colour, but males are generally paler than females. In the centre of the wing-cases (across the suture) there is a pale mark in the shape of a horse-shoe which is always visible, even on the darkest specimens. In pale specimens, the centre of the horse-shoe remains dark, forming a single black spot, and there is usually another V-shaped pale mark below the 'U' of the horse-shoe. The pronotum is the same brown colour as the wing-cases but is pale at the sides. The ladybird has only been found by beating but is easily overlooked on the beating tray because of its small size and close resemblance to similarly-coloured small spiderlings (RGB).

This is such a rarity in Britain that one has to go to continental literature for information. The adult, larva and pupa are all illustrated by Klausnitzer (1997), who also quotes five different authors as all in agreement that the species is a specialist feeder on whitefly (family Aleyrodidae: tiny aphid-like bugs with white wings). The ladybird is a warmth-loving Mediterranean species that has shown signs of spreading northwards in Europe since about 1980, and is most often found by beating old ivy in spring (Fürsch, 1967).

There were only a handful of records from Britain before its recent capture in Surrey. One was found on ivy in Leicestershire in 1872, followed by a few specimens from sites in Berkshire and Oxfordshire in 1915. Breeding colonies were found in Oxford by N.J. Mills in 1979 and 1980, when the ladybird was feeding on a species of whitefly on bushes of *Viburnum tinus*, a common garden shrub (Pope, 1987). The only previous Surrey record was made over one hundred years ago, in 1890, when one was beaten from ivy in Headley Lane near Box Hill by T.H. Hall (Fowler & Donisthorpe, 1913).

A single example was found by Ian Menzies at Bookham Common in February, 1992. This first specimen was beaten from holly under oak trees but its discovery led to a concerted effort to find the species on ivy, and another specimen was found in March by Roger Booth on ivy growing up tall oaks in the same area of the common. Further searches over the next few years produced additional specimens so the species is clearly well-established at Bookham but apparently in low numbers, although most of the ivy is out of reach (Menzies, 1993, 1994, 1996, 1997).

A surprising new development has been the discovery of the species on honeysuckle in another part of Bookham Common in 1999, by Matthew Hogg and Maxwell Barclay. The honeysuckle was growing over a birch tree at the edge of a wood, and several specimens were found on three different occasions.

The contrast in fortunes between the two small, rare ladybirds on ivy is remarkable. This species was discovered at Bookham in 1992 but remains confined to one site, while *Nephus quadrimaculatus* has spread rapidly since 1993 to become a common insect over more than half the county. It seems possible that *Clitostethus* could also occur at other sites in Surrey, so old ivy growing on mature trees, whitefly-infested *Viburnum* and even honeysuckle are all worth examining.

RECORDS: **Bookham Common, South-east Wood** (TQ1255), beaten from holly beneath oak, 29.2.92 (ISM), beaten from ivy on trunks of oak trees, 7.3.92 (RGB), 14.8.92 (ISM), 6.3.93 (RGB), 10.12.94, 11.3.95, 8.4.95 (3 specimens), 21.5.95, 11.10.95 (ISM, RGB), 26.5.96 (M. Collier), 8.6.96 (MVLB); **Bookham Common, Eastern Plain** (TQ1256), three beaten from honeysuckle, 31.8.99 (MVLB), 21.9.99 (3), 18.10.99 (2) (both ISM).

## *Platynaspis luteorubra* (Goeze, 1777)

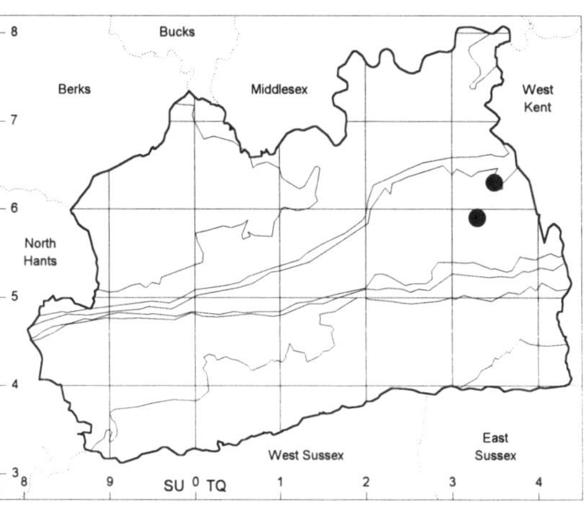


**National Status:** Notable (NA)
**Number of Tetrads:** 2
**Status in Surrey:** Rare
**Habitat:** Dry places, living with ants

This is perhaps the most ladybird-like of all the smaller coccinellids. One which landed on my arm was immediately recognised as a tiny ladybird by a colleague on a conservation working party. Alas, my hands were full of a bundle of alien weeds and the insect flew off before it could be named with absolute certainty, so it has not been included here as a record.

The wing-cases are black with four red spots and covered with downy hairs, so the only possible confusion is with *Nephus quadrimaculatus*. The present species is significantly larger at about 3 mm long and the foremost spot on each wing-case is differently shaped, being roundish with an irregular outline and larger than the other spot. The front angles of the pronotum are almost always yellowish, the broad head is also yellow in the male but black in the female, and the pale tibiae are particularly broad and flattened.

The species has been found at only two localities in Surrey during this survey and we have to go back to the Oxfordshire of over 40 years ago for some insight into its habits. John Pontin was studying the interactions between aphids and ants when he found an unknown brown woodlouse-shaped larva in the underground galleries of the common black ant *Lasius niger*. The larva was feeding on a root aphid and he reared it out to an adult *Platynaspis luteorubra*. He commented on the similarity in shape between this larva and those of blue butterflies and the hoverfly *Microdon*, which are also found in ants' nests, and thought it likely that *Platynaspis* was a normal associate of underground aphids and ants (Pontin, 1960). Michael Majerus has also found a larva feeding on an underground aphid tended by ants, but both Majerus (1994) and Klausnitzer (1997) illustrate it feeding on aphids on the upper stems of thistles. These photographs originate from a study of the species in Germany by W. Völkl, but it is not at all clear if the larva occurs on the aerial stems of plants in the wild as well as in the laboratory.

Our specimen from Croham Hurst was found above ground, by observation, in the small piece of chalk grassland at one end of this hill of sand and pebbles. The three examples from Riddlesdown Quarry resulted from over 100 hours of observing insects at this site in 1997. All came from a small area on the quarry floor with a sparse flora of grasses, composites and trefoils. The first was sieved from leaf litter in a shallow mossy ditch which may have been its overwintering site, in spite of the late date of 19 May. The other two were extremely active and flew readily but seemed to be attracted to the leaf-rosettes of dandelions and other composites. Nests of *Lasius niger* were frequent here, as were other ants such as *Myrmica scabrinodis* and *Formica fusca*.

RECORDS: **Croham Hurst** (TQ3462), 3.9.82 (GBC); **Riddlesdown Quarry** (TQ3359), sieved from leaf litter, 19.5.97, around leaves of yellow composite, 25.5.97, at base of young dandelion, 9.6.97 (RDH).

## *Hyperaspis pseudopustulata* Mulsant, 1853 PLATE 13

**National Status:** Notable (NB)
**Number of Tetrads:** 0
**Status in Surrey:** No recent records
**Habitat:** Various

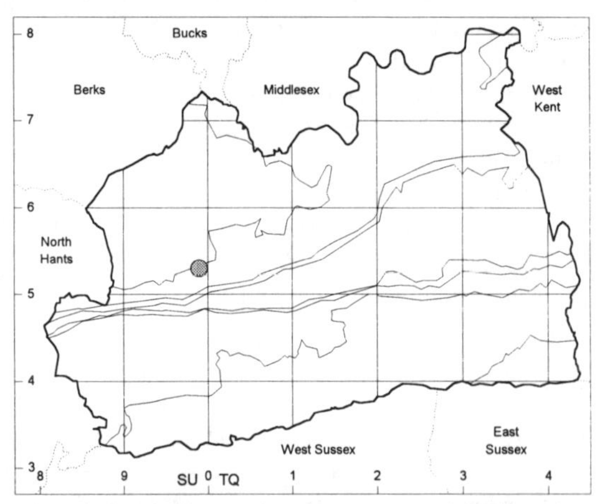


The ancient world had a great many fabulous beasts, from mermaids to unicorns, and belief in some of them persists to this day. Often one is unsure whether or not to believe in their existence. My fabulous creature is *Hyperaspis*, since we have completely failed to find it in Surrey during the 20 years of this survey and I have never seen a specimen. This may just be a matter of not knowing exactly where to look. The main hosts are species of mealybug and other soft scale-insects (Roger Booth, *pers. comm.*) but I have not yet acquired the knack of finding these small creatures.

Our colour plate shows an example loaned to our photographer by courtesy of Michael Majerus and the Cambridge Ladybird Survey. It is a shiny black insect, 3 mm long, with a pair of red spots near the tips of the wing-cases and another pair at the front angles of the pronotum. The rear angles are sharp, so the pronotum and wing-cases form a continuous outline. The antennae are short but, unlike other small ladybirds, this species is completely hairless. The larvae have a white waxy covering and so resemble species of *Scymnus*.

Records from Chobham, Esher, Mickleham, Box Hill and Shirley are listed by Fowler (1889), and it was again recorded from Chobham on 12 May 1906 by G.C. Champion. The only relatively modern record for Surrey is from Whitmoor Common on 28 August 1975 by Peter Hodge.

The name *Hyperaspis reppensis* was used for this species for many years, but Allen (1969) showed that British specimens conform to a different European species, *pseudopustulata*. All *Hyperaspis* are uncommon insects in central Europe (Fürsch, 1967).

# APPENDIX 1 – References

**Allen, A.A., 1953.**
Further notes on *Scymnus* (*Pullus*) *limbatus* Steph. (Col., Coccinellidae). *Entomologist's Mon. Mag.* **89**: 283.

**Allen, A.A., 1969.**
The identity of the British species of *Hyperaspis* Redtb. (Col., Coccinellidae). *Entomologist's Mon. Mag.* **104** (1968): 232.

**Allen, A.A., 1974.**
Some insects of an iron-oxide tip at Greenwich, S.E. *Entomologist's Mon. Mag.* **109** (1973): 235.

**Baldwin, A.J., 1988.**
Biological observations on *Subcoccinella vigintiquattuorpunctata* (L.) (Col., Coccinellidae). *Entomologist's Mon. Mag.* **124**: 57-61.

**Baldwin, A.J., 1990.**
Further biological observations on *Subcoccinella vigintiquattuorpunctata* (L.) (Col., Coccinellidae). *Entomologist's Mon. Mag.* **126**: 223-229.

**Barson, G., & Carter, C.I., 1972.**
A Species of Phylloxeridae, *Moritziella corticalis* (Kalt.) (Homoptera) New to Britain, and a Key to the British Oak-feeding Phylloxeridae. *Entomologist* **105**: 130-134.

**Benham, B.R., & Muggleton, J., 1970.**
Studies on the ecology of *Coccinella undecimpunctata* Linn. (Col. Coccinellidae). *Entomologist* **103**: 153-170.

**Benham, B.R., & Muggleton, J., 1979.**
Observations on the over-wintering of Coccinellidae in the British Isles. *Entomologist's Mon. Mag.* **114** (1978): 191-197.

**Booth, R.G., 1997.**
Review of *Ladybirds* by Michael E.N.Majerus. *The Coleopterist* **6**: 78-79.

**Booth, R.G. (in press).**
[*Rhyzobius lophanthae*. Exhibit at BENHS Annual Exhibition 1999.] *Br. J. Ent. Nat. Hist.* **13**.

**Buck, F.D., 1957.**
*Harmonia quadripunctata* (Pont.) (Col., Coccinellidae) in Surrey. *Entomologist's Mon. Mag.* **93**: 72, 195.

**Burton, J.F., 1968.**
*The Oxford Book of Insects*. Oxford University Press, Oxford.

**Chinery, M., 1986.**
*Collins Guide to the Insects of Britain and Western Europe.* Collins, London.

**Constantine, B., & Majerus, M.E.N., 1994.**
*Cryptolaemus montrouzieri* Mulsant (Col., Coccinellidae) in Britain. *Entomologist's Mon. Mag.* **130**: 45-46.

**Cox, D., 1946.**
Coleoptera at Colchester, Essex. *Entomologist's Mon. Mag.* **82**: 47.

**Crowson, R.A. & E.A., 1956.**
Coleoptera and Hemiptera-Heteroptera in the Silverdale area (North-west Lancashire). *Entomologist's Mon. Mag.* **92**: 230.

**Dandy, J.E., 1969.**
*Watsonian vice-counties of Great Britain.* Ray Society, London.

**Donisthorpe, H.St.J., 1919-20.**
The Myrmecophilous Lady-Bird, *Coccinella distincta* Fald., its Life-history and Association with Ants. *Entomologist's Rec. J. Var.* **31**: 214-222, **32**: 1-3.

**Duffy, E.A.J., 1945.**
The coleopterous fauna of the Hants-Surrey border. *Entomologist's Mon. Mag.* **81**: 169-179.

**Eastop, V.F., & Pope, R.D., 1966.**
Notes on the ecology and phenology of some British Coccinellidae. *Entomologist* **99**: 287-289.

**Eastop, V.F., & Pope, R.D., 1969.**
Notes on the biology of some British Coccinellidae. *Entomologist* **102**: 162-164.

**Fowler, W.W., 1889.**
*The Coleoptera of the British Islands.* Volume 3. Reeve, London.

**Fowler, W.W., & Donisthorpe, H.St.J., 1913.**
*The Coleoptera of the British Islands.* Volume 6. Ballantyne, London.

**Fürsch, H., 1967.**
Coccinellidae *in* Freude, H., Harde, K.W., & Lohse, G.A. (eds.), *Die Käfer Mitteleuropas*, Volume 7: 227-278. Goecke & Evers, Krefeld.

**Geoghegan, I.E., Thomas, W.P., & Majerus, M.E.N., 1997.**
Notes on the coccinellid parasitoid *Dinocampus coccinellae* (Schrank) (Hymenoptera: Braconidae) in Scotland. *Entomologist* **116**: 179-184.

**Goss, H., 1902.**
Coleoptera. *A History of the County of Surrey*, **3** *Zoology*. Constable, London.

**Greathead, D.J., 1973.**
A review of introductions of *Lindorus lophanthae* (Blaisd.) (Col: Coccinellidae) against hard scales (Diaspididae). *Commonwealth Institute of Biological Control, Technical Bulletin*, Number 16: 29-33.

**Halstead, A.J., 1999.**
Frost survival by *Cryptolaemus montrouzieri* Muls. (Coleoptera: Coccinellidae) in an outdoor situation. *Br. J. Ent. Nat. Hist.* **11** (1998): 163-164.

**Hammond, P.M., 1985.**
Dimorphism of wings, wing-folding and wing-toileting devices in the ladybird, *Rhyzobius litura* (F.) (Coleoptera: Coccinellidae), with a discussion of inter-population variation in this and other wing-dimorphic beetle species. *Biol. J. Linn. Soc.* **24**: 15-33.

**Hawkins, R.D. (in press).**
*Rhyzobius chrysomeloides* (Herbst) (Coleoptera, Coccinellidae) new to Britain. *Br. J. Ent. Nat. Hist.* **13**.

**Heal, N.F., 1994.**
[*Nephus quadrimaculatus*. Exhibit at BENHS Annual Exhibition 1993.] *Br. J. Ent. Nat. Hist.* **7**: 170.

**Hodek, I., 1973.**
*Biology of Coccinellidae.* Junk, The Hague, and Academia, Prague.

**Hodge, P.J. (in press).**
[*Rhyzobius chrysomeloides*. Exhibit at BENHS Annual Exhibition 1999.] *Br. J. Ent. Nat. Hist.* **13**.

**Horion, A., 1961.**
*Faunistik der mitteleuropäischen Käfer*, Volume 8. Feyel, Überlingen.

**Hyman, P.S., 1992.**
*A Review of the Scarce and Threatened Coleoptera of Great Britain* (part 1). Joint Nature Conservation Committee, Peterborough.

**Iablokoff-Khnzorian, S.M., 1982.**
*Les Coccinelles. Coléoptères-Coccinellidae. Tribu Coccinellini des régions Paléarctique et Orientale*. Boubée, Paris.

**Jones, R.A., 1999a.**
Aggregation of over one million 16-spot ladybirds in a bramble hedge, and "blushing" in two specimens. *Br. J. Ent. Nat. Hist.* **12**: 89-91.

**Jones, R.A., 1999b.**
Entomological surveys of vertical river flood defence walls in urban London – brownfield corridors: problems, practicalities and some promising results. *Br. J. Ent. Nat. Hist.* **12**: 193-213.

**Joy, N.H., 1932.**
*A Practical Handbook of British Beetles* (2 vols.). Witherby, London.

**Klausnitzer, B. & H., 1997.**
*Marienkäfer*, 4th edition. Westarp, Magdeburg.

**Klausnitzer, B., & Bellmann, C., 1969.**
Zum Vorkommen von Coccinellidenlarven (Coleoptera) in Bodenfallen auf Fichtenstandorten. *Ent. Nachr.* **13**: 128-132 [Ladybird larvae in pitfall traps under spruce].

**Lousley, J.E., 1976.**
*Flora of Surrey*. David & Charles, Newton Abbot.

**Majerus, M.E.N., 1988.**
Some notes on the 18-spot ladybird (*Myrrha 18-guttata* L.) (Coleoptera: Coccinellidae). *Br. J. Ent. Nat. Hist.* **1**: 11-13.

**Majerus, M.E.N., 1989.**
*Coccinella magnifica* (Redtenbacher): a myrmecophilous ladybird. *Br. J. Ent. Nat. Hist.* **2**: 97-106.

**Majerus, M.E.N., 1991.**
Habitat and host plant preferences of British ladybirds (Col., Coccinellidae). *Entomologist's Mon. Mag.* **127**: 167-175.

**Majerus, M.E.N., 1994.**
*Ladybirds*. The New Naturalist series, 81. Harper Collins, London.

**Majerus, M.E.N., 1997.**
Parasitization of British ladybirds by *Dinocampus coccinellae* (Schrank) (Hymenoptera: Braconidae).
*Br. J. Ent. Nat. Hist.* **10**: 15-24.

**Majerus, M.E.N., & Kearns, P.W.E., 1989.**
*Ladybirds*. Naturalists' Handbooks 10. Richmond Publishing, Slough.

**Majerus, M.E.N., Majerus, T.M.O., Bertrand, D., & Walker, L.E., 1997.**
The geographic distributions of ladybirds (Coleoptera: Coccinellidae) in Britain (1984-1994).
*Entomologist's Mon. Mag.* **133**: 181-203.

**Majerus, M.E.N., & Williams, Z., 1989.**
The distribution and life history of the Orange Ladybird, *Halyzia sedecimguttata* (L.) (Coleoptera: Coccinellidae) in Britain. *Entomologist's Gazette* **40**: 71-78.

**Marriner, T.F., 1927.**
Observations on the life history of *Subcoccinella 24-punctata.*
*Entomologist's Mon. Mag.* **63**: 118-123.

**Menzies, I.S., 1993.**
Coleoptera *in* Survey of Bookham Common, Progress Report for 1992. *London Naturalist* **72**: 108-109.

**Menzies, I.S., 1994.**
[*Clitostethus arcuatus*. Exhibit at BENHS Annual Exhibition 1993.] *Br. J. Ent. Nat. Hist.* **7**: 172.

**Menzies, I.S., 1996.**
Beetles *in* Survey of Bookham Common, Progress Report for 1995. *London Naturalist* **75**: 128-131.

**Menzies, I.S., 1997.**
Beetles *in* Survey of Bookham Common, Progress Report for 1996. *London Naturalist* **76**: 183-184.

**Menzies, I.S., 1999.**
[*Rhyzobius chrysomeloides*. Exhibit at BENHS Annual Exhibition 1998.] *Br. J. Ent. Nat. Hist.* **12**: 176.

**Menzies, I.S., & Spooner, B.M., 2000.**
*Henosepilachna argus* (Geoffroy) (Coccinellidae, Epilachninae), a phytophagous ladybird new to the U.K., breeding at Molesey, Surrey. *The Coleopterist* **9**: 1-4.

**Merritt Hawkes, O.A., 1927.**
*Coccinella 10-punctata* L. – a tromorphic [= trimorphic] ladybird. *Entomologist's Mon. Mag.* **63**: 203-208.

**Mills, N.J., 1981.**
Essential and alternative foods for some British Coccinellidae (Coleoptera). *Entomologist's Gazette* **32**: 197-202.

**Moon, A., 1986.**
*Ladybirds in Dorset*. Dorset Environmental Records Centre, Dorchester.

**Newstead, R., 1901, 1903.**
*Monograph of the Coccidae of the British Isles*. Volumes 1 & 2. Ray Society, London.

**Owen, J.A., 1993.**
[*Nephus quadrimaculatus*. Exhibit at BENHS Annual Exhibition 1992.] Br. J. Ent. Nat. Hist. **6**: 76.

**Pontin, A.J., 1960.**
Some records of predators and parasites adapted to attack aphids attended by ants. *Entomologist's Mon. Mag.* **95** (1959): 154-155.

**Pope, R.D., 1953.**
Coccinellidae and Sphindidae. *Handbooks for the identification of British Insects* **5(7)**. Royal Entomological Society, London.

**Pope, R.D., 1973.**
The species of *Scymnus* (s.str.), *Scymnus* (*Pullus*) and *Nephus* (Col., Coccinellidae) occurring in the British Isles. *Entomologist's Mon. Mag.* **109**: 3-39.

**Pope, R.D., 1977.**
Brachyptery and wing-polymorphism among the Coccinellidae. *Systematic Entomology* **2**: 59-66.

**Pope, R.D., 1987.**
*Clitostethus arcuatus* and *Nephus quadrimaculatus*, *in* Shirt, D.B. (ed.), *British Red Data Books: 2. Insects*. Nature Conservancy Council, Peterborough.

**Richards, A.M., Pope, R.D., & Eastop, V.F., 1976.**
Observations on the biology of *Subcoccinella vigintiquattuorpunctata* (L.) in southern England. *Ecol. Ent.* **1**: 201-207.

**Robbins, S., 2000.**
Some notes of ladybirds in North Cornwall. *Bull. Amat. Ent. Soc.* **59**: 21-27.

**Rotheray, G.E., 1989.**
*Aphid Predators*. Naturalists' Handbooks 11. Richmond Publishing, Slough.

**Sloggett, J.J., Manica, A., Day, M.J., & Majerus, M.E.N., 1999.**
Predation of ladybirds (Coleoptera: Coccinellidae) by wood ants, *Formica rufa* L. (Hymenoptera: Formicidae). *Entomologist's Gazette* **50**: 217-221.

**Stace, C.A., 1997.**
*New Flora of the British Isles*. Second edition. Cambridge University Press, Cambridge.

**Stroyan, H.L.G., 1984.**
Aphids - Pterocommatinae and Aphidinae (Aphidini). *Handbooks for the identification of British Insects* **2(6)**. Royal Entomological Society, London.

**Tanasijevic, N., 1958.**
Zur Morphologie und Biologie des Luzernemarienkäfers *Subcoccinella vigintiquatuorpunctata* (L.) (Col., Coccinellidae). *Beitr. Ent.* **8**: 23-78.

**Van Emden, F.I., 1949.**
Larvae of British Beetles. VII. (Coccinellidae). *Entomologist's Mon. Mag.* **85**: 265-283.

**Verdcourt, B., 1952.**
Further records of Bedfordshire Coleoptera. *Entomologist's Mon. Mag.* **88**: 73-80.

**Ward, L.K., 1970.**
*Aleuropteryx juniperi* Ohm (Neur., Coniopterygidae) new to Britain feeding on *Carulaspis juniperi* Bouché (Hem., Diaspididae). *Entomologist's Mon. Mag.* **106**: 74-78.

**Ward, L.K., 1977.**
The conservation of juniper: the associated fauna with special reference to southern England. *J. appl. Ecol.* **14**: 81-120.

# APPENDIX 2 – Gazetteer of sites

| Site | Grid reference |
|---|---|
| **Addiscombe** | TQ3467 |
| **Ashtead Common** | TQ1759 |
| **Bagmoor Common** | SU9242 |
| **Bagshot** | SU9163 |
| **Bagshot Heath** | SU8963 |
| **Basingstoke Canal** | SU8851-TQ0561 |
| **Battersea Park** | TQ2877 |
| **Blackheath, near Guildford** | TQ0346 |
| **Bookham Common** | TQ1256 |
| **Box Hill** | TQ1751 |
| **Brentmoor Heath** | SU9361 |
| **Brockham Quarry** | TQ2051 |
| **Brockwell Park** | TQ3174 |
| **Buckland Hills** | TQ2252 |
| **Burpham** | TQ0152 |
| **Camberwell** | TQ3377 |
| **Capel** | TQ1740 |
| **Caterham** | TQ3455 |
| **Charlwood** | TQ2441 |
| **Chiddingfold** | SU9635 |
| **Chilworth** | TQ0247 |
| **Chobham Common** | SU9963 |
| **Clark's Green** | TQ1739 |
| **Claygate** | TQ1563 |
| **Cobham** | TQ1059 |
| **Cockshot Wood, Mickleham** | TQ1853 |
| **Coldharbour** | TQ1443 |
| **Coulsdon** | TQ3058 |
| **Cranleigh** | TQ0639 |
| **Croydon** | TQ3265 |
| **Dawcombe** | TQ2152 |
| **Dorking** | TQ1649 |
| **Earlswood** | TQ2749 |
| **Epsom Common (Stew Ponds)** | TQ1860 |
| **Esher Common** | TQ1262 |
| **Ewhurst** | TQ0940 |
| **Farnham** | SU8446 |
| **Farthing Downs** | TQ3057 |
| **Field Common** | TQ1367 |
| **Gatwick Airport** | TQ2841 |
| **Godalming** | SU9643 |
| **Guildford** | SU9949 |
| **Hackhurst Downs** | TQ0948 |
| **Halfway Lane, Godalming** | SU9543 |
| **Hankley Common** | SU8841 |
| **Haslemere** | SU9032 |
| **Haxted** | TQ4145 |
| **Headley Heath** | TQ2053 |
| **Hedgecourt Pond** | TQ3540 |
| **Horley** | TQ2843 |
| **Horley (Balcombe Road)** | TQ2942 |
| **Horsell Common** | TQ0060 |
| **Hound House Farm, Newdigate** | TQ2043 |
| **Inholms Lane, North Holmwood** | TQ1747 |
| **Juniper Hall, Mickleham** | TQ1752 |
| **Kew** | TQ1977 |
| **Kew Gardens** | TQ1876 |
| **Kingswood** | TQ2456 |

| | |
|---|---|
| **Leatherhead** | TQ1656 |
| **Leith Hill** | TQ1343 |
| **Lightwater Country Park** | SU9162 |
| **Limpsfield Chart** | TQ4352 |
| **London** | TQ17/27/37 |
| | |
| **Merrow Downs** | TQ0249 |
| **Merstham** | TQ2953 |
| **Mickleham** | TQ1753 |
| **Mitcham Common** | TQ2967 |
| **Molesey** | TQ1468 |
| **Molesey Heath** | TQ1367 |
| **Morden Park** | TQ2467 |
| | |
| **Newdigate** | TQ1942 |
| **Newdigate Brickworks** | TQ2042 |
| **Newlands Corner** | TQ0449 |
| **Nonsuch Park** | TQ2363 |
| **Norbury Park** | TQ1553 |
| **North Holmwood** | TQ1647 |
| **Norwood** | TQ3270 |
| **Nower Wood** | TQ1954 |
| **Nunhead Cemetery** | TQ3575 |
| **Nutfield** | TQ3050 |
| | |
| **Oxshott** | TQ1460 |
| **Oxshott Heath** | TQ1461 |
| | |
| **Park Downs, Banstead** | TQ2658 |
| **Petersham Meadows** | TQ1773 |
| **Putney** | TQ2375 |
| **Puttenham Common** | SU9146 |
| | |
| **Ranmore Common** | TQ1250 |
| **Reigate** | TQ2550 |
| **Richmond** | TQ1875 |
| **Riddlesdown** | TQ3260 |
| **Rowhill Copse** | SU8549 |
| | |
| **Salfords** | TQ2846 |
| **Sheets Heath** | SU9457 |
| **Shirley** | TQ3564 |
| **Smallfield** | TQ3143 |
| **Somersbury Wood** | TQ1037 |
| **South Croydon** | TQ3363 |
| **Sutton (Brighton Road)** | TQ2562 |
| | |
| **Thursley Common** | SU9040 |
| | |
| **Vann Lake** | TQ1539 |
| | |
| **Waddon** | TQ3165 |
| **Walton-on-Thames** | TQ1066 |
| **Wandsworth Common** | TQ2774 |
| **Wey Navigation** | SU9744-TQ0765 |
| **Whitmoor Common** | SU9853 |
| **Wimbledon Common** | TQ2372 |
| **Winterfold Forest** | TQ0642 |
| **Wisley Common** | TQ0758 |
| **Wisley RHS Garden** | TQ0658 |
| **Witley Common** | SU9341 |
| **Woking** | TQ0058 |
| **Woodlands Park** | TQ1458 |
| **Wormley** | SU9438 |

# APPENDIX 3 – Glossary

**adelgids** — the woolly aphids of coniferous trees.

**elytra** — the wing-cases of an adult beetle.

**femur** (plural **femora**) — the first long section of the leg of an insect.

**FSC** — Field Studies Council: a body providing environmental education and other courses for biology students and adults.

**genus** — a group of closely-related species, corresponding to the first word of the scientific name of an insect or other organism.

**Malaise trap** — fixed arrangement of nets for sampling flying insects (see page 19).

**metasternum** — the rear section of the lower part of the thorax.

**postcoxal lines** — semicircular grooves or ridges on the underside of the abdomen behind the base of each hind leg.

**pronotum** — the front section of the upper part of the thorax, i.e. the visible part of the thorax as seen from above, lying between the head and the wing-cases.

**prosternal keels** — two low ridges between the bases of the front legs.

**psyllids** — small insects related to the leaf-hoppers, also known as jumping plant-lice.

**scutellum** — a tiny triangular plate at the base of the wing-cases; actually part of the thorax.

**suture** — the line along which the wing-cases meet.

**tarsus** (plural **tarsi**) — the many-jointed foot of an insect.

**tetrad** — a square of 2 km by 2 km, represented by a single dot on the distribution maps.

**tibia** (plural **tibiae**) — the second long section of the leg of an insect.

**vice-county** — a county (or part of a county) with unalterable boundary, used for biological recording since the 1850's (see page 7).

**WEA** — Workers' Educational Association: an independent organisation providing evening and daytime classes for adults.

# INDEX – Plants, with scientific names

## INDEX – Plants, with scientific names (continued)

## INDEX – Plants, with scientific names (continued)

# INDEX – Ladybirds (scientific names)

Figures in bold indicate plate numbers

# INDEX – Ladybirds (English names)

Figures in bold indicate plate numbers

# INDEX – Other insects and food

Figures in bold indicate plate numbers